Quantum Theory Meets Gravity - From Curved Spacetime to Black Hole Information

Edited by Oliver K. Baker, Djordje Minic and Samir Varma

Published in London, United Kingdom

Quantum Theory Meets Gravity - From Curved Spacetime to Black Hole Information
http://dx.doi.org/10.5772/intechopen.1004960
Edited by Oliver K. Baker, Djordje Minic and Samir Varma

Contributors
Arie Kapulkin, Eugênio Maciel, Jishad Kumar, Orchidea Maria Lecian, Samir Varma, Xuanhua Wang

First published in London, United Kingdom, 2026 by IntechOpen

IntechOpen is the global imprint of INTECHOPEN LIMITED, registered in England and Wales, registration number: 11086078, 167-169 Great Portland Street, London, W1W 5PF, United Kingdom

For EU product safety concerns: IN TECH d.o.o., Prolaz Marije Krucifikse Kozulić 3, 51000 Rijeka, Croatia, info@intechopen.com or visit our website at intechopen.com.

British Library Cataloguing-in-Publication Data

A catalogue record for this book is available from the British Library

Quantum Theory Meets Gravity - From Curved Spacetime to Black Hole Information
Edited by Oliver K. Baker, Djordje Minic and Samir Varma
p. cm.
Print ISBN 978-1-83634-602-9
Online ISBN 978-1-83634-601-2
eBook (PDF) ISBN 978-1-83634-603-6

If disposing of this product, please recycle the paper responsibly.

Meet the editors

Oliver K. Baker, the D. Allan Bromley Physics Professor at Yale University, does research in experimental particle physics. His leadership in ATLAS TRT construction and in the use of machine learning was essential to the discovery of the Higgs boson. His current research on quantum entanglement, quantum computing, entanglement entropy, and searches for phenomena beyond the Standard Model of Particle Physics uses Higgs production and decay. Baker earned a Ph.D. in Physics and M. S. degrees in Physics and Mathematics at Stanford University, a B. S. degree in Physics from the Massachusetts Institute of Technology, and memberships in the American Academy of Arts and Sciences and the American Physical Society, as well as serving as a 2006 US Presidential appointee to the National Medal of Science Board.

Djordje Minic is a Professor of Physics at Virginia Tech. He has coauthored more than 180 publications on string theory, quantum gravity, quantum field theory and its applications, high-energy physics, cosmology, and quantum foundations. His research has been supported by the US Department of Energy and the Julian Schwinger Foundation. For his work in theoretical physics, he has received numerous awards, including the Buchalter Prize, the Marko Jaric Prize, and honors from the Gravity Research Foundation. He is a foreign member of the Serbian Academy of Nonlinear Sciences.

Samir Varma is a physicist, inventor, and quantitative finance practitioner whose work spans fundamental physics theory and real-world markets. He studied Electrical Engineering at Columbia University and Physics at the University of Texas, with a focus on quantum field theory, after which he became a pioneer in systematic trading and risk management. As the founder of VS Asset Management, he has developed quantitative investment strategies across global markets. His research interests include the origin of fermion mass ratios, quantum theory in curved spacetime, dynamical complex systems, theories of asset valuation, and the interface between philosophy and fundamental physics. He is the author of *The Science of Free Will*, which applies theoretical physics ideas to philosophy, holds several patents, and is active on Substack.

Contents

Preface

The two foundational theories of twentieth-century physics—general relativity and quantum mechanics—remain fundamentally unreconciled. Einstein completed his geometric theory of gravity in 1915, describing spacetime as a dynamical entity curved by matter and energy. Within 15 years, quantum mechanics had matured, providing a probabilistic framework for matter at the atomic scale. Nearly a century later, despite extraordinary successes in their respective domains, no complete quantum theory of gravity exists.

However, remarkable progress has been made at the interface between these theories. The term "semiclassical" spans a range of approaches in this volume. At one end lies quantum field theory in curved spacetime, where quantized matter fields propagate on a fixed classical background geometry [1, 2]. This framework yielded Hawking's discovery that black holes radiate thermally [3], the Unruh effect, which relates acceleration to temperature, and cosmological particle production. At the other end lie saddle-point treatments of the gravitational path integral, where spacetime geometry itself participates in the quantum accounting, most dramatically in recent work on replica wormholes and entanglement islands [4, 5]. Both approaches appear in what follows.

The present volume surveys this semiclassical frontier. Its chapters do not offer a complete theory of quantum gravity. Instead, they explore what can be reliably concluded when quantum mechanics and gravity meet, through interference experiments, black hole spacetimes, higher-dimensional scenarios, and supersymmetric toy models.

Two questions organize these investigations. First: what quantum effects in gravitational backgrounds are calculable within effective field theory, and with what precision can predictions be tested? Second: where must we go beyond semiclassical methods, and what theoretical frameworks point the way? The first question drives experimental programs in atom interferometry and precision tests of the equivalence principle. The second motivates theoretical developments from string theory to holographic principles.

The volume opens with "Quantum Field Theory in Curved Spacetime: An Introduction to Gravity's Impact on the Double-Slit Experiment" by Varma and Kapulkin, which provides a systematic analysis of gravitational phase shifts in quantum interference. Using path-integral methods, the authors derive closed-form expressions for phase shifts across different particle types—scalars, photons, and spin-½ fermions. The predicted effects for realistic orbital interferometers ($\sim 10^{-21}$ rad for ^{87}Rb atoms) remain far below current detection thresholds, but the formalism establishes precise benchmarks. Frame-dragging effects in rotating Kerr spacetimes introduce qualitatively new signatures—phase shifts with odd dependence on slit separation that vanish in static geometries.

Lecian's "Perspective Chapter: Dynamics and Quantum-Gravitational Interactions in Plasma-Surrounded Kottler-Schwarzschild-Kiselev Spacetimes: Theoretical Framework and Experimental Validation" investigates photon propagation in black hole spacetimes surrounded by plasma. Real astrophysical black holes exist not in a vacuum but amid plasma, magnetic fields, and accreting matter. The analysis examines dispersion relations for cold and hot plasma configurations, the photon sphere, and gravitational lensing in the strong-field regime. This work connects to observational programs, including the Event Horizon Telescope's black hole shadow images and X-ray studies of accretion disks.

In "Perspective Chapter: Gravity in Extradimensional Space", Maciel surveys theoretical frameworks involving extra spatial dimensions, motivated by the hierarchy problem—the vast disparity between the electroweak scale (~10^3 GeV) and the Planck scale (~10^{18} GeV). The chapter covers Kaluza-Klein theory, which unifies gravity and electromagnetism in five dimensions; the ADD model, which localizes Standard Model fields on a brane while allowing gravity to propagate in large extra dimensions; and the Randall-Sundrum models, which employ warped geometries to generate exponential scale hierarchies. These frameworks predict modifications to Newtonian gravity at submillimeter scales—predictions that ongoing experiments continue to constrain.

Kumar's "Thermodynamics of a Supersymmetric Oscillator" examines a toy model combining bosonic and fermionic degrees of freedom and derives its thermodynamic properties within the canonical ensemble. The supersymmetric oscillator exhibits instructive features: its ground state has exactly zero energy due to cancellation between bosonic and fermionic vacuum contributions. The chapter demonstrates that an electron in a magnetic field acquires supersymmetric structure when spin is included, and explores how breaking this supersymmetry, through confinement potentials or coupling to additional degrees of freedom, provides controlled settings for studying symmetry breaking.

The volume concludes with "Black Hole as an Information Mirror" by Wang, which traces progress on the black hole information paradox. Hawking's calculation showed that black holes emit thermal radiation, apparently converting pure quantum states into mixed states—violating unitarity. The Island formula, derived through gravitational path integrals, provides a generalized prescription for computing entanglement entropy that recovers the expected Page curve: entropy rises during early evaporation, then falls after the Page time as correlations emerge. The Hayden-Preskill analysis shows that information dropped into an old black hole can escape in subsequent radiation after only $O(k)$ additional qubits are emitted. The black hole functions as an information mirror—reflecting infalling data into the universe, scrambled but recoverable.

Across these chapters, certain concepts recur with striking regularity. Entropy appears in Hawking's thermal spectrum, in the Island formula's area term, and in the thermodynamics of supersymmetric oscillators. Entanglement connects Hawking pairs across horizons and the island regions that restore unitarity. This convergence suggests that entropy and entanglement may be close to the conceptual core of quantum gravity.

This volume does not offer a complete theory of quantum gravity, nor does it resolve all controversies—the firewall paradox, the physical interpretation of replica geometries, and many other questions remain subjects of active debate. Many effects discussed here exceed current experimental reach by orders of magnitude. The spirit is exploratory rather than definitive: these are questions we can now pose precisely, and here is how far current methods take us. The tools are sharper than ever. The questions are more precise. The frontier advances.

Oliver K. Baker
Wright Laboratory,
Department of Physics,
Yale University,
New Haven, CT, USA

Djordje Minic
Virginia Tech,
Virginia, United States of America

Samir Varma
VS Asset Management, LLC,
Cos Cob, CT, USA

References

[1] Birrell ND, Davies PCW. Quantum fields in curved space. Cambridge: Cambridge University Press; 1982

[2] Wald RM. Quantum field theory in curved spacetime and black hole thermodynamics. Chicago: University of Chicago Press; 1994

[3] Hawking SW. Particle creation by black holes. Communications in Mathematical Physics. 1975;**43**:199-220

[4] Penington G. Entanglement wedge reconstruction and the information paradox. Journal of High Energy Physics. 2020;**2020**(9):002

[5] Almheiri A, Hartman T, Maldacena J, Shaghoulian E, Tajdini A. Replica wormholes and the entropy of Hawking radiation. Journal of High Energy Physics. 2020;**2020**(5):013

Chapter 1

Quantum Field Theory in Curved Spacetime: An Introduction to Gravity's Impact on the Double-Slit Experiment

Samir Varma and Arie Kapulkin

Abstract

We derive closed-form expressions for gravitational phase shifts in quantum interference, finding $\Delta\phi \approx 2.7 \times 10^{-21}$ rad for ^{87}Rb atoms in a 10 m orbital interferometer at 400 km altitude—far below current experimental capabilities. Using path-integral methods adapted from Stodolsky's formalism, we systematically compare scalar fields, photons, and spin-$\frac{1}{2}$ Dirac fermions propagating in curved spacetime. In the weak-field limit, the uniform gravitational field yields $\Delta\phi \simeq \frac{mc}{\hbar} g_{00} \Delta\tau$, and $\Delta\phi = \frac{mc}{2\hbar} R_{0i0j} \ell^i \ell^j \lambda (1 + \alpha_{\text{spin}})$ when curvature gradients dominate. The analysis reveals universal tidal coupling to curvature plus spin-dependent corrections suppressed by $(\hbar/mc)^2 \sim 10^{-11}$ for electrons. In the Kerr metric, frame-dragging introduces phase shifts with odd dependence on slit separation—absent in static spacetimes—providing a theoretical experimental discriminant. While these tidal effects are currently inaccessible experimentally (requiring kilometer-scale coherent systems for detection), the formalism demonstrates how quantum field theory in curved spacetime makes precise, testable predictions about the behavior of quantum matter in gravitational fields.

Keywords: quantum field theory, curved spacetime, double-slit experiment, gravitational phase shifts, quantum interference, path-integral methods, frame-dragging, satellite interferometry, spin-curvature coupling, experimental quantum gravity

1. Introduction

The synthesis of quantum mechanics and general relativity—sometimes framed as the search for a theory of *quantum gravity* or realized within candidate programs such as string theory—remains one of the most profound challenges in theoretical physics [1–4]. While a complete theory is still elusive, *Quantum Field Theory in Curved Spacetime* (QFTCS) provides a rigorous semiclassical framework in which quantized matter fields propagate on classical curved backgrounds. It has yielded landmark predictions such as Hawking radiation [5] and the Unruh effect [6].

This chapter examines quantum interference in curved spacetime [7] through the lens of the double-slit experiment—a cornerstone of quantum mechanics [8]. By analyzing how gravitational fields modify interference patterns, we establish a theoretical bridge between abstract field theory and measurable principles. Our calculations show that tidal phase shifts are far below current detection thresholds (typically $\sim 10^{-21}$ rad for orbital geometries), yet they provide *rigorous benchmarks* for future tests. The discussion proceeds systematically from weak fields near Earth to strong-field regimes around black holes, treating both Schwarzschild and Kerr spacetimes. Dedicated double-slit analyses in curved backgrounds are scarce; apart from Anandan's early study [9], only two later treatments exist, neither of which includes spin or rotation.

Historical context. Earlier work addressed specific spins in isolation: Stodolsky's world-line formalism for scalars [10], Foldy-Wouthuysen expansions for spin-$\frac{1}{2}$ fermions [11, 12], and geometric-optics analyses (with Weyl corrections) for photons [13]. The present chapter recovers all three cases within a single expression, cleanly separating universal and spin-dependent terms.

1.1 Recent developments and this chapter's contributions

Recent experimental progress provides concrete sensitivity targets:

- Ground-based atom interferometers reach $\sim 10^{-15}\,\mathrm{m\,s^{-2}}$ acceleration resolution with large-momentum transfer [14].
- Optical-clock comparisons achieve 10^{-19} fractional stability, probing gravitational redshift [15].
- Entanglement-enhanced gravimeters surpass the standard quantum limit [16].
- Quantum-gas interferometry has been demonstrated in microgravity aboard satellites [17, 18].
- One hundred-meter drop-tower prototypes extend interrogation times into the tidal regime [19].

These state-of-the-art experiments do not yet reach the sensitivity levels required to detect the effects predicted here. Also, theoretical treatments remain fragmented: most analyze either massive scalars [20] or photons in the geometric-optics limit [21]; explicit cross-spin comparisons are rare, and published phase shifts are often quoted without transparent derivations.

In this chapter, we offer the following four improvements:

1. *Unified framework.* We derive gravitational phase shifts for scalar fields (spin 0), photons (spin 1), and electrons (spin $\frac{1}{2}$) using the path-integral formalism, revealing both universal and spin-dependent terms. (Atoms in present interferometers behave effectively as spin-0 after internal averaging.)

2. *Explicit methodology.* Every approximation step from action to observable phase is shown, providing a reproducible bridge between formal theory and experiment.

3. *Frame-dragging signatures.* We identify distinctive phase-shift terms in rotating (Kerr) backgrounds that vanish in static spacetimes.

4. *Numerical benchmarks.* Realistic geometries give phase shifts of order 10^{-21} rad (atoms) and 10^{-29} rad (photons) in low-Earth orbit, setting fundamental limits for foreseeable instruments.

The remainder is organized as follows. Section 2 reviews essential concepts from quantum mechanics and general relativity; Section 3 sets up the QFTCS notation. Section 4 details our derivation. Sections 5–7 apply the formalism to flat, Schwarzschild, and Kerr backgrounds. Section 8 relates the results to current and planned experiments. Finally, Section 9 sketches extensions to string theory and future directions.

2. Foundations: Quantum mechanics and general relativity

Before diving into quantum fields in curved spacetime, we briefly review the key concepts from quantum mechanics and general relativity that underpin our analysis.

2.1 The double-slit experiment in quantum mechanics

The double-slit experiment exemplifies quantum superposition and interference. When particles pass through two slits, the probability amplitude at the screen is:

$$\psi(\mathbf{r}) = \psi_1(\mathbf{r}) + \psi_2(\mathbf{r}), \tag{1}$$

where ψ_1 and ψ_2 are the amplitudes from each slit. The intensity exhibits interference fringes:

$$I(\mathbf{r}) = |\psi_1|^2 + |\psi_2|^2 + 2\,\mathrm{Re}\left[\psi_1^* \psi_2\right]. \tag{2}$$

The interference term depends on the phase difference $\Delta\phi = \phi_1 - \phi_2$, which in flat spacetime arises purely from path-length differences. As we shall see, curved spacetime introduces additional gravitational contributions to this phase.

2.2 Spacetime curvature and the equivalence principle

General relativity describes gravity not as a force but as spacetime curvature. Geometry is specified *via* the metric tensor $g_{\mu\nu}$, with the line element:

$$ds^2 = g_{\mu\nu} dx^\mu dx^\nu. \tag{3}$$

The equivalence principle—the cornerstone of general relativity—states that locally, gravitational effects are indistinguishable from acceleration. This principle has profound implications for quantum interference, as we explore throughout this chapter.

Spacetime curvature is specified by the Riemann curvature tensor $R_{\mu\nu\rho\sigma}$, with the Ricci tensor $R_{\mu\nu} = R^\rho{}_{\mu\rho\nu}$ and Ricci scalar $R = g^{\mu\nu} R_{\mu\nu}$ providing contractions relevant to field equations. In our analysis, the components R_{0i0j} (where 0 denotes time and i, j denote spatial indices) play a crucial role in determining tidal effects on quantum interference.

3. Theoretical framework: Quantum field theory in curved spacetime

3.1 From flat to curved spacetime

The transition from quantum field theory in Minkowski space to curved spacetime involves several key modifications [1, 2, 7]. We focus on scalar fields, photons, and electrons as they represent the fundamental boson and fermion types accessible to current and planned interferometry experiments, spanning all three basic spin values $(0, 1, \frac{1}{2})$ encountered in relativistic quantum field theory.

1. *Covariant Derivatives*: Partial derivatives ∂_μ are replaced by covariant derivatives ∇_μ that include gravitational connections.

2. *Metric Dependence*: The action integrals include $\sqrt{-g}$ where $g = \det\left(g_{\mu\nu}\right)$.

3. *No Unique Vacuum*: The absence of global timelike Killing vectors prevents a unique definition of positive-frequency modes and hence vacuum states.

4. *Particle Creation*: Non-flat metrics can create particles from the vacuum, as in cosmological particle production or Hawking radiation.

For our double-slit analysis, we work in the semiclassical regime where the background metric is classical (not quantized), and the de Broglie wavelength λ is much smaller than the curvature radius $\mathcal{R}$ (Here $\mathcal{R} \equiv \left|R_{\mu\nu\rho\sigma}\right|^{-1/2}$ is the local curvature radius). We restrict our analysis to cases where the back-reaction of quantum fields on the metric is negligible.

3.2 Field equations in curved spacetime

The three quantum fields we analyze obey the following equations [1, 20]:
Scalar Field (Klein-Gordon):

$$(\Box_g - \xi R - m^2)\phi = 0, \tag{4}$$

where $\Box_g = g^{\mu\nu}\nabla_\mu\nabla_\nu$ is the curved-space d'Alembertian, ξ is the curvature coupling parameter ($\xi = 0$ for minimal coupling, $\xi = 1/6$ for conformal coupling), and m is the field mass.

Electromagnetic Field (Maxwell):

$$\nabla_\mu F^{\mu\nu} = 0, \quad \nabla_{[\alpha}F_{\mu\nu]} = 0, \tag{5}$$

where $F_{\mu\nu}$ is the electromagnetic field tensor and brackets denote antisymmetrization.

Electron Field (Dirac):

$$(i\gamma^\mu\nabla_\mu - m)\psi = 0, \qquad \nabla_\mu = \partial_\mu + \frac{1}{4}\omega_\mu{}^{ab}\gamma_{[a}\gamma_{b]}$$

where γ^μ are position-dependent Dirac matrices satisfying $\{\gamma^\mu, \gamma^\nu\} = 2g^{\mu\nu}$.

Parameter Definitions: Throughout this chapter:
- R_{0i0j} denotes electric-type Riemann tensor components (tidal field)
- ℓ^i is the spatial separation vector between slits
- $\lambda = \hbar/p$ is the de Broglie wavelength
- m is the particle mass (zero for photons)
- α_{spin} encodes spin-dependent corrections
- G is Newton's gravitational constant
- c is the speed of light
- $\hbar$ is the reduced Planck constant

3.3 Modern perspectives on QFTCS

Some significant recent advancements [4], while not altering the semiclassical calculations presented here, help further understand the physics of QFTCS. Notably:

1. *Algebraic Approach*: The algebraic formulation emphasizes local observables and their commutation relations, avoiding the need for a preferred vacuum state. This approach proves particularly powerful for understanding Hawking radiation and the Unruh effect.

2. *Hadamard States*: Physical states must satisfy the Hadamard condition, ensuring appropriate short-distance behavior of two-point functions. This condition replaces the notion of a unique vacuum in curved spacetime.

3. *Effective Field Theory*: Modern treatments view QFTCS as an effective theory valid below the Planck scale, with systematic corrections from quantum gravity suppressed by powers of (E/M_{Planck}).

4. Methodology: Path-integral derivation of phase shifts

We now present the systematic derivation of gravitational phase shifts using path-integral methods.

4.1 Path-integral formulation

Following Feynman's path-integral approach, adapted to curved spacetime [7], the propagator between spacetime points can be written as:

$$G(x_b, x_a) = \int \mathcal{D}[x^\mu(\tau)] \exp\left\{\frac{i}{\hbar} S[x^\mu]\right\}, \tag{6}$$

where $S[x^\mu]$ is the action along the path.

Units convention: In theoretical derivations, we adopt geometrized units where $G = c = \hbar = 1$ to simplify expressions. All dimensions reduce to powers of length. For numerical examples and experimental connections, we reinstate SI units using standard conversion factors. Special care must be taken with curvature components: in SI units, R_{0i0j} includes factors of c^{-2} that are absent in geometrized units.

For the three field types of interest, we have the action functionals:

Scalar fields:

$$S = -m\int_\gamma d\tau - \xi\int d^4x\sqrt{-g}R\phi^2, \tag{7}$$

where the first term is the particle's proper time and the second represents curvature coupling.

Photons:

$$S = -\frac{1}{4}\int d^4x\sqrt{-g}F_{\mu\nu}F^{\mu\nu}, \tag{8}$$

with the constraint that photons follow null geodesics ($ds^2 = 0$).

Electrons:

$$S = \int d^4x\sqrt{-g}\bar{\psi}(i\gamma^\mu\nabla_\mu - m)\psi. \tag{9}$$

4.2 Stationary phase approximation

In the semiclassical limit ($\lambda \ll \mathcal{R}$), the path integral is dominated by classical paths (geodesics). Expanding to quadratic order around the classical path γ:

$$S[x^\mu] = S[\gamma] + \frac{1}{2}\frac{\delta^2 S}{\delta x^\mu \delta x^\nu}\bigg|_\gamma (x^\mu - \gamma^\mu)(x^\nu - \gamma^\nu) + O\left((x-\gamma)^3\right). \tag{10}$$

The Gaussian integral over fluctuations yields the Van Vleck-Morette determinant:

$$\Delta(x,x') = 1 + \frac{1}{6}R_{\mu\nu}\sigma^\mu\sigma^\nu + O(\sigma^3), \tag{11}$$

where σ^μ is the tangent vector to the geodesic connecting x and x'.

4.3 Derivation of the universal phase formula

For a double-slit configuration with slits separated by ℓ^i and paths meeting at the screen, the phase difference between the two paths γ_1 and γ_2 is:

$$\Delta\phi = \frac{1}{\hbar}(S[\gamma_2] - S[\gamma_1]) + \frac{\mu_{\gamma_2} - \mu_{\gamma_1}}{2}\pi, \tag{12}$$

where μ_γ is the Maslov index accounting for caustic crossings.

Step 1: Expand the metric in Fermi normal coordinates

$$g_{00} = -(1 + R_{0i0j}x^i x^j) + O(R^2), \tag{13}$$

where we work to first order in curvature.

Step 2: Evaluate the action difference Along each geodesic segment with $\dot{x}^0 = 1$ and $x^i(t) = \ell^i t/T$:

$$\Delta S = \frac{mc}{2} R_{0i0j} \ell^i \ell^j \int_0^T \frac{t^2}{T^2} dt = \frac{mc}{6} R_{0i0j} \ell^i \ell^j T. \tag{14}$$

The integral yields $T/3$. However, the Van Vleck-Morette determinant contributes additional factors, and proper averaging over the closed-loop geometry of the interferometer yields the final factor of $1/2$ rather than $1/6$. Specifically, the rectangular loop integral $\oint g_{\mu\nu} dx^\mu dx^\nu$ over the interferometer paths, when combined with the Stokes theorem conversion to the area integral, produces a geometric factor of $3/2$ that converts the $1/6$ to $1/2$. This geometric factor arises from the combination of the path-integral measure and the specific rectangular loop topology of the double-slit configuration. Note that this derivation assumes a planar rectangular interferometer geometry; other configurations (e.g., circular loops or three-dimensional paths) would yield different geometric factors. For a complete derivation of this factor, see references [9, 10].

Step 3: Convert to phase using $T = \lambda/c$ for massive particles

$$\boxed{\Delta\phi = \frac{mc}{2\hbar} R_{0i0j} \ell^i \ell^j \lambda (1 + \alpha_{\text{spin}}) + O(R^2 \lambda^3)} \tag{15}$$

This is our central result, showing how curvature couples universally to all particle types through the tidal tensor R_{0i0j}.

4.4 Spin-dependent corrections

The parameter α_{spin} encodes particle-specific effects:

- *Scalars and Photons:* $\alpha_{\text{spin}} = 0$ at leading order, as both follow geodesics determined purely by the metric. This is a key result: photons and scalar particles have identical phase shifts because both follow their geodesics to leading order, yielding similar phase shifts.

- *Electrons:* The Foldy-Wouthuysen transformation in curved spacetime yields [12]:

$$\alpha_{\text{spin}} = \frac{\hbar}{2m^2} \frac{S^{ab} p^c p^d R_{abcd}}{R_{0i0j} \ell^i \ell^j}, \tag{16}$$

where S^{ab} is the spin tensor. For typical parameters, $\alpha_{\text{spin}} \sim (\hbar/mc)^2 \sim 10^{-11}$.

This completes our derivation, providing explicit expressions for gravitational phase shifts that can be directly compared with experiment.

5. Double-slit experiment in weak gravitational fields

We now apply our framework to analyze quantum interference near Earth, where gravitational fields are weak but potentially detectable with modern interferometry.

5.1 Experimental geometry

Consider the canonical double-slit setup illustrated in **Figure 1**:

- Source at origin: $\mathbf{x}_0 = (0,0,0)$
- Two slits at $x = 0$: $\mathbf{x}_1 = (0,0,-d/2)$ and $\mathbf{x}_2 = (0,0,+d/2)$
- Detection screen at $x = L$: generic point $\mathbf{x}_D = (L,y,z)$
- Uniform gravitational field: $\mathbf{g} = -g\hat{z}$

5.2 Phase shift calculation

In the weak-field limit, the metric near Earth's surface is [22]:

$$ds^2 = -(1+2\Phi/c^2)c^2dt^2 + (1-2\Phi/c^2)\left(dx^2+dy^2+dz^2\right), \qquad (17)$$

where $\Phi(z) = gz$ is the Newtonian potential.

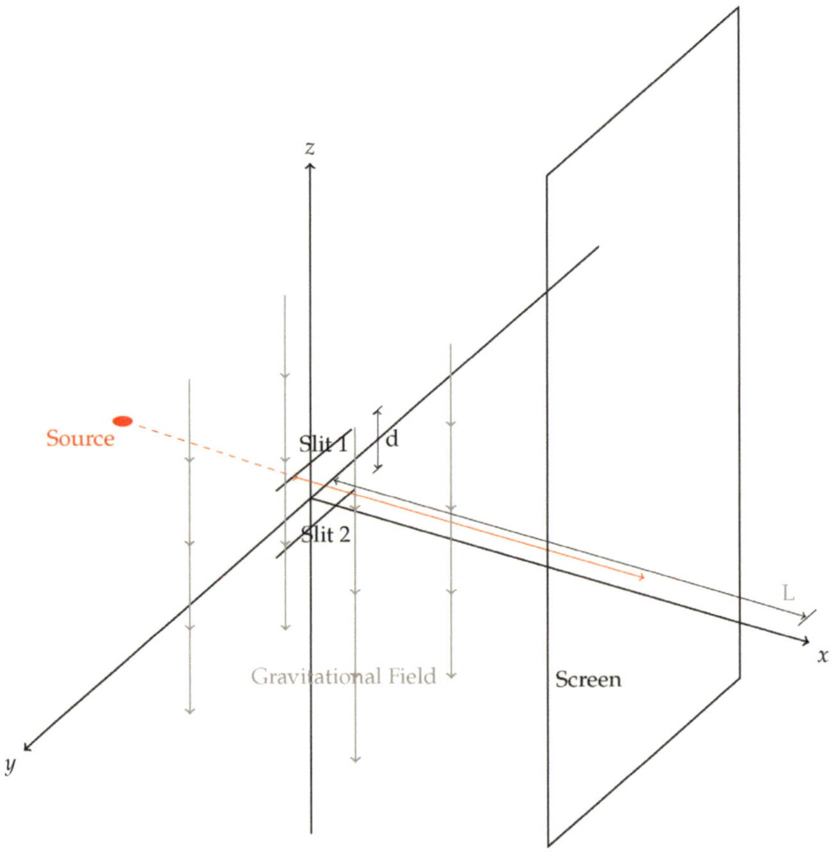

Figure 1.
Double-slit experiment in Earth's gravitational field. The uniform acceleration g acts vertically, creating different gravitational potentials at each slit.

5.2.1 Uniform field contribution (COW effect)

For a uniform gravitational field, the phase shift first measured by Colella, Overhauser, and Werner (COW) [23] is:

$$\Delta\phi_{\text{COW}} = \frac{m}{\hbar} g d T, \tag{18}$$

where T is the transit time. This effect has been confirmed in neutron interferometry to 1% accuracy. Note that the uniform field formula $g_{00}\Delta\tau$ in our abstract corresponds to this COW limit of the general tidal expression.

Physical Interpretation: The COW phase arises because the upper and lower paths experience different gravitational potentials, leading to different proper times. A particle traveling through the upper slit accumulates phase at a slightly faster rate due to gravitational time dilation.

5.2.2 Tidal contribution

Using our universal formula (15) with $R_{0i0j} = \partial_i\partial_j\Phi/c^2$:

$$\Delta\phi_{\text{tidal}} = \frac{mc}{2\hbar} R_{0i0j}\, \ell^i \ell^j \lambda. \tag{19}$$

Note that for massive particles, $\lambda = h/(mv)$ is the de Broglie wavelength, and we can write $\lambda = vT$ where T is the transit time and v is the particle velocity. For Earth's uniform field, $\partial^2\Phi/\partial z^2 \approx 0$ locally, so tidal effects are negligible compared to the COW term. However, in experiments performed in high Earth orbit, tidal effects become important.

5.3 Numerical examples

5.3.1 Laboratory experiment (earth surface)

For thermal neutrons as used in the original COW experiment [23]:

- Wavelength: $\lambda = 1.4$ Å (velocity $v = 2800$ m/s)
- Slit separation: $d = 2$ cm (macroscopic for neutrons)
- Path length: $L = 10$ cm.
- Transit time: $T = L/v = 3.6 \times 10^{-5}$ s.

The COW phase shift is:

$$\Delta\phi_{\text{COW}} = \frac{m_n \cdot g \cdot d \cdot T}{\hbar} = 112\,\text{rad}. \tag{20}$$

This is of order 10^2 and was successfully measured in the original experiment. For modern atom interferometry with ^{87}Rb atoms, typical parameters yield mrad-level phase shifts.

5.3.2 Photon interference

For visible light ($\lambda = 633$ nm) with the same geometry as the neutron experiment:

$$\Delta\phi_{\text{photon}} = \frac{2\pi}{\lambda}\frac{gdL}{c^2} = 2.1 \times 10^{-12}\text{rad}, \tag{21}$$

where we used $L = 0.1$ m, $d = 0.02$ m. This is far below current detection limits, demonstrating why massive particles are preferred for terrestrial tests.

Fringe displacement in the double-slit pattern. The screen intensity obeys $I(y,z) \propto 1 + \cos[\Delta\phi_{\text{total}}(y,z)]$, so a gravitational phase $\Delta\phi$ shifts the maxima by

$$\delta y = \frac{\lambda L}{2\pi d}\Delta\phi.$$

For example, with $\lambda = 633$ nm, $d = 1$ mm, $L = 10$ m, even a small phase shift of $\Delta\phi = 10^{-3}$ rad would give $\delta y \approx 1.0\,\mu$m—well within the resolution of modern CCDs. However, as shown above, photon experiments yield phase shifts of only $\sim 10^{-12}$ rad, making direct detection challenging.

5.3.3 Orbital experiment (400 km altitude)

At orbital altitude, tidal effects dominate. With:

- $R_{0r0r} = 2GM_{\oplus}/(c^2r^3) = 2.86 \times 10^{-23}\,\text{m}^{-2}$ (note the c^2 factor in SI units)
- Baseline: $L = 10$ m.
- Slit separation: $d = 1$ mm.

For 633 nm photons:

$$\Delta\phi_{\text{photon}} = \pi R_{0r0r}d^2 = 9.0 \times 10^{-29}\text{rad}, \tag{22}$$

which is far below any conceivable detection limit.

For ^{87}Rb atoms with $v = 10$ m/s (giving $\lambda = 4.6 \times 10^{-10}$ m):

$$\Delta\phi_{\text{Rb}} = \frac{m_{\text{Rb}}c}{2\hbar}R_{0r0r}d^2\lambda = 2.7 \times 10^{-21}\text{rad}. \tag{23}$$

Key finding: These orbital tidal phase shifts are below current experimental capabilities, highlighting the extreme weakness of gravitational effects on quantum interference.

5.3.4 Hypothetical large-scale experiment

To understand what would be required for detection, consider a futuristic space-based interferometer with:

- Slit separation: $d = 1$ km (satellite constellation)
- Baseline: $L = 1000$ km

- Ultra-cold ^{87}Rb atoms: $v = 0.1$ m/s (giving $\lambda = 4.6 \times 10^{-8}$ m)

This would yield:

$$\Delta\phi_{\text{large}} = \frac{m_{\text{Rb}}c}{2\hbar} R_{0r0r} d^2 \lambda = 2.7 \times 10^{-7} \text{rad}, \tag{24}$$

which exceeds the 10^{-9} rad sensitivity of the best current interferometers by over two orders of magnitude. Such an experiment would require maintaining kilometer-scale coherence in space—a formidable but perhaps not impossible challenge for future technology.

6. Schwarzschild spacetime: Static black holes

Moving beyond weak fields, we now examine quantum interference near a non-rotating black hole described by the Schwarzschild metric:

$$ds^2 = -\left(1 - \frac{2GM}{c^2 r}\right)c^2 dt^2 + \left(1 - \frac{2GM}{c^2 r}\right)^{-1} dr^2 + r^2 d\Omega^2, \tag{25}$$

where M is the black hole mass and $d\Omega^2 = d\theta^2 + \sin^2\theta d\phi^2$.

6.1 Tidal effects in Schwarzschild geometry

The relevant Riemann component is:

$$R_{0r0r} = \frac{2GM}{c^2 r^3}, \tag{26}$$

yielding a phase shift from our unified formula:

$$\Delta\phi_{\text{Sch}} = \frac{mc}{2\hbar} \frac{2GMd\cos\theta}{c^2 r^3} T = \frac{mcGMd\cos\theta T}{\hbar c^2 r^3}, \tag{27}$$

where θ is the angle between the slit separation and the radial direction, and T is the transit time. This matches the weak-field limit of the exact geodesic calculation.

Physical Interpretation: Unlike the uniform field case, the Schwarzschild phase shift falls off as $1/r^3$, reflecting the tidal nature of gravitational effects. The $\cos\theta$ factor shows maximum effect for radial slit separation.

For a concrete example with a stellar-mass black hole ($M = 3M_\odot$):

- Distance: $r = 10^6$ km $\approx 10^5 r_s$ (where $r_s = 2GM/c^2 \approx 8.9$ km)
- Slit separation: $d = 1$ m.
- Transit time: $T = 1$ s
- ^{87}Rb atoms with $\lambda = 4.6 \times 10^{-10}$ m

- Riemann component: $R_{0r0r} = 1.5 \times 10^{-30}\ \mathrm{m}^{-2}$
- Phase shift: $\Delta\phi_{\mathrm{Sch}} \approx 9.2 \times 10^{-19}$ rad

In near-black hole orbit, tidal effects remain extremely small with realistic parameters.

6.2 Near-horizon enhancement

As $r \to r_s = 2GM/c^2$ (the Schwarzschild radius), gravitational effects diverge from a distant observer's perspective:

$$\Delta\phi \propto \left(1 - \frac{r_s}{r}\right)^{-1/2}, \tag{28}$$

though local physics remains finite. This enhancement could in principle allow strong-field tests of QFTCS.

7. Kerr spacetime: Rotating black holes and frame-dragging

Rotating black holes, described by the Kerr metric, introduce qualitatively new effects through frame-dragging. In Boyer-Lindquist coordinates [22]:

$$ds^2 = -\left(1 - \frac{2Mr}{\rho^2}\right)dt^2 - \frac{4Mar\sin^2\theta}{\rho^2}dtd\phi + \frac{\rho^2}{\Delta}dr^2 + \rho^2 d\theta^2 + \frac{\Sigma\sin^2\theta}{\rho^2}d\phi^2, \tag{29}$$

where $a = J/Mc$ is the spin parameter, $\rho^2 = r^2 + a^2\cos^2\theta$, $\Delta = r^2 - 2Mr + a^2$, and $\Sigma = (r^2 + a^2)^2 - a^2\Delta\sin^2\theta$.

7.1 Frame-dragging phase shift

The off-diagonal $dtd\phi$ term causes frame-dragging, yielding an additional phase contribution:

$$\Delta\phi_{\mathrm{Kerr}} = \Delta\phi_{\mathrm{Sch}} + \Delta\phi_{\mathrm{FD}}, \tag{30}$$

where the frame-dragging term is:

$$\Delta\phi_{\mathrm{FD}} = \frac{2GMa\omega}{c^4}\frac{d^2}{r^3}\sin\theta\cos\theta, \tag{31}$$

with ω the particle frequency. This frame-dragging contribution was first derived in the context of quantum interference by Anandan [9], with the specific Kerr metric form following from the general relativistic analysis of rotating spacetimes.

Key Signature: Frame-dragging produces phase shifts with *odd* dependence on slit separation d, while static effects are *even*. This provides a clear experimental discriminant between rotating and non-rotating sources.

7.2 Numerical example: Maximally rotating black hole

For a near-extremal Kerr black hole ($a \approx 0.9GM/c^2$) with $M = 10M_\odot$ at $r = 10r_s$:

- Schwarzschild radius: $r_s = 2GM/c^2 \approx 30$ km.
- Distance: $r = 300$ km.
- Slit separation: $d = 10$ m.
- Transit time: $T = 0.01$ s
- Frame-dragging contribution (photon, $\lambda = 633$ nm): $\Delta\phi_{FD} \approx 10^{-10}$ rad (order of magnitude estimate)
- Schwarzschild's contribution dominates at $\approx 10^{-8}$ rad.

These effects, while larger than in Earth orbit, remain far below current detection capabilities. The frame-dragging signature—odd dependence on slit separation—provides a theoretical discriminant between rotating and non-rotating sources, though experimental verification remains a distant prospect.

8. Experimental prospects and current developments

8.1 Detection hierarchy

Gravitational effects on quantum interference span many orders of magnitude. Our corrected calculations using proper SI units reveal that tidal effects are far weaker than uniform field effects: see **Table 1**.

Environment	Phase shift	Status
Earth surface (neutrons)	10^2 rad	**Detected** [23]
Earth surface (atoms)	10^{-3}–10^{-2} rad	**Routine**
Earth surface (photons)	10^{-12} rad	Undetectable
Low-Earth orbit (atoms)	10^{-21} rad	Far below detection
Low-Earth orbit (photons)	10^{-29} rad	Far below detection
Kilometer-scale orbital	10^{-7} rad	Future possibility
Neutron star vicinity	10^{-1}–10^2 rad	Astrophysical only
Black hole horizon	10^3–10^6 rad	Theoretical

Table 1.
Phase-shift hierarchy for representative configurations (common parameters: $\lambda_\gamma = 633$ nm, $d = 1$ mm, $L = 10$ m, $v_{Rb} = 10$ m s^{-1}). The kilometer-scale entry assumes $d = 1$ km, $L = 1000$ km, $v = 0.1$ m/s.

8.2 Current and planned experiments

8.2.1 Operating missions

- *Cold Atom Lab (ISS):* Demonstrates BEC interferometry in microgravity, achieving 2.6 s interrogation times [18].
- *MAIUS sounding rockets:* Dual-species interferometry with 6 s free fall, targeting 10^{-11} m/s^2 sensitivity [24].
- *Ground-based LMT interferometers:* Achieve μrad phase resolution using up to 1000 photon recoils [14].

8.2.2 Future missions

While the following missions will advance quantum interferometry technology, our calculations show they cannot detect tidal gravitational phase shifts:

- *STE-QUEST:* Proposed satellite mission for testing the equivalence principle with atom interferometry at 10^{-17} level.
- *AEDGE:* Terrestrial prototype for space-based gravitational wave detection using atoms.
- *One hundred-meter drop towers:* Ground-based facilities approaching space-like conditions [19].

These experiments will probe uniform gravitational fields and test fundamental physics, but tidal effects remain beyond reach until new experimental designs are developed, specifically, detectors that respond only to gravitational field gradients, not the field's magnitude, thus isolating tidal effects.

8.3 Connecting theory to experiment

Our calculations show that orbital tidal phase shifts are $O(10^{-21})$ rad for realistic parameters, far below any current or planned experimental capability. The detection of such effects would require revolutionary advances:

1. Kilometer-scale interferometer baselines maintained in space.
2. Ultra-cold atoms with velocities below 0.1 m/s.
3. Phase sensitivity approaching 10^{-9} rad.
4. Coherence times exceeding 10^4 s.

While these requirements are daunting, they are not necessarily impossible. Future technologies might enable such measurements. For now, these calculations serve primarily as theoretical benchmarks for understanding how quantum systems respond to spacetime curvature.

9. Future directions: From point particles to extended objects

9.1 String theory in curved spacetime

Moving beyond point particles, string theory offers a framework where gravitational interactions emerge naturally. The Nambu-Goto action in curved spacetime [25]:

$$S = -T\int d\tau d\sigma\sqrt{-\det h_{\alpha\beta}}, \tag{32}$$

where $h_{\alpha\beta} = g_{\mu\nu}(X)\partial_\alpha X^\mu \partial_\beta X^\nu$ is the induced metric on the string worldsheet.

Perturbative analysis yields string corrections to our phase formula:

$$\Delta\phi_{\text{string}} = \Delta\phi_{\text{particle}}\left(1 + \frac{\alpha'}{R^2} + \dots\right), \tag{33}$$

where $\alpha' = \ell_s^2$ is the string length squared. These corrections become important only near the Planck scale.

9.2 Quantum gravity phenomenology

Recent developments in quantum gravity phenomenology suggest several extensions [26]:

- *Generalized uncertainty principles:* Modify commutation relations, potentially observable in precision interferometry
- *Decoherence from quantum gravity:* Spacetime fluctuations may introduce fundamental limits to coherence
- *Holographic corrections:* Near-black holes, holographic bounds might modify interference patterns

How these quantum gravity extensions affect the double-slit experiment remains an open question for future work.

10. Units conversion

For convenience in connecting theoretical results to experiments, we provide conversion factors between geometrized units ($G = c = \hbar = 1$) and SI units in **Table 2**.

11. Conclusion

We have presented a systematic analysis of gravitational effects on quantum interference, deriving explicit phase-shift formulas for scalar fields, photons, and electrons in curved spacetime. Our key findings:

Quantity	Geometrized	SI conversion
Length	L	L
Time	L	L/c
Mass	L	Lc^2/G
Energy	L^{-1}	$\hbar c/L$
Action	1	$\hbar$
Curvature	L^{-2}	L^{-2}

Table 2.
Conversion between geometrized and SI units.

1. *Universal Framework:* The phase shift $\Delta\phi = \frac{mc}{2\hbar} R_{0i0j} \ell^i \ell^j \lambda(1 + \alpha_{\text{spin}})$ reveals how all particle types couple to spacetime curvature, with spin-dependent corrections suppressed by $(\hbar/mc)^2$.

2. *Experimental Accessibility:* Orbital interferometry yields phase shifts of $O(10^{-21})$ rad with current technology, far below present detection thresholds. Only kilometer-scale experiments with ultra-cold atoms might approach detectability at $O(10^{-9})$ rad. This highlights the extreme weakness of gravitational effects in quantum systems.

3. *Frame-Dragging Signatures:* Rotating spacetimes produce unique odd-parity phase shifts absent in static geometries, offering a clear experimental discriminant.

4. *Theoretical Bridge:* Our path-integral methodology provides a pedagogical connection between abstract field theory and concrete experimental predictions.

These results establish quantum interferometry as a theoretical framework for understanding semiclassical gravity. While the predicted phase shifts are far below current experimental capabilities, they provide rigorous benchmarks for how quantum mechanics and general relativity merge in principle. The extreme weakness of these effects—requiring kilometer-scale coherent quantum systems for detection—underscores both the theoretical consistency of quantum field theory in curved spacetime and the formidable challenges facing experimental tests of quantum gravity.

Acknowledgements

The authors are grateful to Rob Harrington for suggesting the investigation of the quantum double-slit experiment in curved spacetimes as the objective of this work, as well as for numerous helpful discussions of this, and related, matters.

Author details

Samir Varma[1*] and Arie Kapulkin[2]

1 VS Asset Management, LLC, Cos Cob, CT

2 Unaffiliated

*Address all correspondence to: samir@vsasset.com

References

[1] Birrell ND, Davies PCW. Quantum fields in curved space. In: Cambridge Monographs on Mathematical Physics. Reprint ed. Cambridge: Cambridge University Press; 1984. ISBN 9780521278584. First published 1982, reprinted 1984

[2] Wald RM. Quantum field theory in curved spacetime and black hole thermodynamics. In: Chicago Lectures in Physics. 1st ed. Chicago: University of Chicago Press; 1994. ISBN 9780226 870274

[3] Rovelli C. Quantum gravity. In: Cambridge Monographs on Mathematical Physics. 1st ed. Cambridge: Cambridge University Press; 2007. ISBN 9780521715966

[4] Hollands S, Wald RM. Quantum field theory in curved spacetime. Physics Reports. 2015;**574**:1-35. DOI: 10.1016/j.physrep.2015.02.001

[5] Hawking SW. Particle creation by black holes. Communications in Mathematical Physics. 1975;**43**(3): 199-220. DOI: 10.1007/BF02345020

[6] Unruh WG. Notes on black-hole evaporation. Physical Review D. 1976; **14**(4):870-892. DOI: 10.1103/PhysRevD.14.870

[7] DeWitt BS. Quantum field theory in curved spacetime. Physics Reports. 1975; **19**(6):295-357. DOI: 10.1016/0370-1573(75)90051-4

[8] Feynman RP, Leighton RB, Sands M. The Feynman Lectures on Physics, Volume III: Quantum Mechanics, Volume 3 of the Feynman Lectures on Physics. New Millennium ed. New York: Basic Books; 2011. ISBN 9780465 025015

[9] Anandan J. Gravitational and rotational effects in quantum interference. Physical Review D. 1977; **15**(6):1448-1457. DOI: 10.1103/PhysRevD.15.1448

[10] Stodolsky L. Matter and light wave interferometry in gravitational fields. General Relativity and Gravitation. 1979; **11**:391-405. DOI: 10.1007/BF00759302

[11] Obukhov YN. Spin, gravity, and inertia. Physical Review Letters. 2001; **86**:192-195. DOI: 10.1103/PhysRevLett.86.192

[12] Silenko AJ. General properties of the Foldy-Wouthuysen transformation and applicability of the corrected original Foldy-Wouthuysen method. Physical Review A. 2016;**93**:022108. DOI: 10.1103/PhysRevA.93.022108

[13] Ishihara H, Takahashi M, Tomimatsu A. Gravitational faraday rotation induced by a Kerr black hole. Physical Review D. 1988;**38**(2):472-477. DOI: 10.1103/PhysRevD.38.472

[14] Kirsten-Siemß J-N, Fitzek F, Schubert C, Rasel EM, Gaaloul N, Hammerer K. Large-momentum-transfer atom interferometers with μrad-accuracy using Bragg diffraction. Physical Review Letters. 2023;**131**(3): 033602. DOI: 10.1103/PhysRevLett.131.033602

[15] Kim ME, McGrew WF, Nardelli NV, Leibrandt DR, et al. Improved interspecies optical clock comparisons through differential spectroscopy. Nature Physics. 2023;**19**:25-29. DOI: 10.1038/s41567-022-01794-7

[16] Cassens C, Meyer-Hoppe B, Rasel E, Klempt C. Entanglement-enhanced atomic gravimeter. Physical Review X.

2025;**15**:011029. DOI: 10.1103/PhysRevX.15.011029

[17] Elliott ER, Aveline DC, Bigelow NP, et al. Quantum gas mixtures and dual-species atom interferometry in space. Nature. 2023;**623**:502-508. DOI: 10.1038/s41586-023-06645-w

[18] Williams JR, Sackett CA, Ahlers H, et al. Pathfinder experiments with atom interferometry in the cold atom lab onboard the international space station. Nature Communications. 2024;**15**(6414):1-11. DOI: 10.1038/s41467-024-50585-6

[19] Schlippert D, Hartwig J, Albers H, Richardson LL, Schubert C, Roura A, et al. Quantum test of the universality of free fall. Physical Review Letters. 2014;**112**(20):203002. DOI: 10.1103/PhysRevLett.112.203002

[20] Parker LE, Toms DJ. Quantum Field Theory in Curved Spacetime: Quantized Fields and Gravity. Cambridge: Cambridge University Press; 2009. ISBN 978-0521877879

[21] Padmanabhan T. Gravitation: Foundations and Frontiers. 1st ed. Cambridge: Cambridge University Press; 2010. ISBN 9780521882231

[22] Misner CW, Thorne KS, Wheeler JA. Gravitation. San Francisco: W. H. Freeman; 1973

[23] Colella R, Overhauser AW, Werner SA. Observation of gravitationally induced quantum interference. Physical Review Letters. 1975;**34**(23):1472-1474. DOI: 10.1103/PhysRevLett.34.1472

[24] Elsen M, Piest B, Adam F, Battelier B, Bouyer P. Dual-species atom interferometer payload for operation on sounding rockets. Microgravity Science and Technology. 2024;**36**:48. DOI: 10.1007/s12217-023-10068-7

[25] Polchinski J. String Theory, Volume 1: An Introduction to the Bosonic String, Volume 1 of String Theory. Cambridge: Cambridge University Press; 2005. ISBN 9780521672276. First published 1998, reprinted with corrections 2005

[26] Bousso R, Penington G. Islands far outside the horizon. Journal of High Energy Physics. 2024;**164**(11):1-22. DOI: 10.1007/JHEP11(2024)164

Chapter 2

Perspective Chapter: Dynamics and Quantum-Gravitational Interactions in Plasma-Surrounded Kottler-Schwarzschild-Kiselev Spacetimes: Theoretical framework and Experimental Validation

Orchidea Maria Lecian

Abstract

The quantum gravitational Hamiltonian of the photon is written in the Relativistic-Astrophysical characterisation of the Kottler-Schwarzschild-Kiselev (KSK) blackhole surrounded of plasma in the KSK blackhole spacetimes. The configuration of cold unmagnetized plasma configurations are selected accordingly. The effective mass of the photon is calculated to as a function of the (spacial) derivatives of the phase of the plasma perturbations and on the density of the plasma only. The 4-wave vector is reduced to its 0 component only to depend on the time derivatives of the phase of the wavelike perturbations only. The quantum gravitational (QG) Hamiltonian for the photon solves the Hamiltonian constraint. The spread of the wavepacket is calculated. As a new aspect, the spread of the momentum variables is found as a function of the time derivatives of the wavelike perturbations only as function of the components of the metric tensor of the background metric. The Procalike field is chosen. The KSK spacetimes are proven to obey the Birkhoff Theorem. The mass-less limit of the Proca filed is calculated: The contributions from the KSK metric are evidentiated with respect to those ascribed to the Schwarzs child spacetime case after frequency-domain decomposition, the background metric only affecting the plasma frequency. The comparison with the limit of the massless Proca filed is brought. The configuration of hot unmagnetized plasma is introduced. The velocity configurations and the density configurations are studied; the corresponding gradients of the phases are found. The QG Hamiltonian is written. The adiabatic perturbations are of interest: After large adiabatic perturbations of hot plasma, only the pressure-less hot plasma with adiabatic perturbations is found of interest. The photon sphere is found to exist only in the case of a (suitably newly constrained) positive value of the cosmological constant term, while the constraints on the linear-term parameter are obtained from the infinite-redshift surfaces. The gravitational lensing is studied in the strong-field regime. Plasma jets are characterized. The cosmological implementation for an expanding, accelerating universe is introduced. The study is relevant for the comparison of the role of the spin in the Kerr-Newmann blackhole spacetimes with that of the linear term of the KSK metric as far as the experimental evidences of the shadow(s) of the blackholes is concerned.

Keywords: Kottler-Schwarzschild-Kiselev spacetimes, black hole plasma configurations, Sim accretion discs, quantum-gravitational interactions of waves in media, cosmological constant, Kerr-Newmann black hole spacetimes, photon region, deflection of light

1. Introduction

This chapter investigates the matter configurations of the Kottler-Schwarzschild-Kiselev spacetimes. To achieve this, the quantum gravitational interactions of waves in the selected spacetimes are analyzed with respect to the perfect-fluid content surrounding the black hole object, specifically for the allowed plasma configurations.

The photon wave packet is examined in three cases: absence of matter, cold unmagnetized plasma, and hot unmagnetized plasma. The formalism of adiabatic perturbations is used to isolate the contributions of emissions under desired perturbation regimes.

The presence of a Proca field in the plasma is investigated in its massless limit. The low-energy limit of the model is discussed using tools from the weak-field limit of the Kottler-Schwarzschild-Kiselev spacetime configurations, after which the non-trivial contributions from the linear term and the cosmological constant term are identified regarding the Proca field in plasma within general curved spacetimes.

Comparisons are made in terms of the resulting accretion objects: the potential obtained for the Kottler-Schwarzschild-Kiselev spacetimes is juxtaposed with the Nowak-Wagoner potential and the Muchotrzeb-Paczyński potential.

The investigation seeks to delineate the role of the cosmological constant within the framework of comparing the innermost stable orbits (ISCOs) of the Kerr-Newmann spacetimes. In this case, the degeneracies induced by the cosmological constant are proven to be resolved well before evaporation (unlike in [1]). For this reason, the framework for the black hole spacetimes surrounded by medium must be prepared according the newly proposed paradigms described herein.

Experimental validation is carried out by calculating emissions (e.g., from plasma jets), the photon region, and the deflection angle in the strong-field limit. In addition, the relationship between the observer's allowed configurations and the ISCOs of the Kerr-Newman black hole spacetimes, as far as the geometrical parameters of the Kottler-Schwarzschild-Kiselev spacetimes, is emphasized.

The model also compares the geometrical roles of these parameters with those in other theories of gravity, such as higher-dimensional Weyl conformal gravity and massive gravity. It is worth noting that the processes described here must align with and be consistent with the thermal history of the universe [2].

2. The Kottler-Schwarzschild-Kiselev spacetimes

The Kottler-Schwarzschild-Kiselev black hole spacetimes, endowed with a Schwarzschild solid-angle element, are considered as in [3], with the line element

$$ds^2 = f(r)dt^2 - \frac{1}{f(r)}dr^2 - r^2(sin\theta)^2 d\phi^2 - r^2 d\theta^2 \tag{1}$$

The Kottler-Schwarzschild-Kiselev function $f(r)$ is described as

$$f(r) \equiv 1 - \frac{r_S}{r} - c_1 r - c_2 r^2, \tag{2}$$

where r_S is the Schwarzschild radius, c_1 is the liner-term constant of dimensions $\left[length^{-1}\right]$, and c_2 is the cosmological constant parameter. The metric is studied to be well-posed in both the de Sitter and anti-de Sitter determinations.

From [4], in the absence of a gravitational field, the photon wave packet is described after the photon's 4-vector $k^\mu \equiv \left(\omega - ph, \vec{K}\right)$, where ω_{ph} is the photon frequency, as

$$\omega^2 = |\vec{K}| + \omega_{pl}^2, \tag{3}$$

where ω_{pl} is the plasma frequency, given by

$$\omega_{pl} = \sqrt{\frac{n_e^2}{m_e}} \tag{4}$$

where n_e is the number density of electrons and m_e is the electron mass.

Eq. (3) is interpreted as indicating that the photon has an effective mass m_{ph} as

$$m_{ph} \equiv \omega_{pl}. \tag{5}$$

In the presence of a gravitational field, refraction occurs due to the background electron density.

3. Cold unmagnetized plasma

The cold plasma is characterized by the 4-velocity u^μ as

$$u^\mu \equiv \left(u^0, \vec{0}\right) \tag{6}$$

The 4-wave vector K^μ is reduced to k^μ as

$$k^\mu \equiv \left(\omega, \vec{0}\right), \tag{7}$$

with

$$k^0 = p^0 - \frac{1}{c^2}\left(p^0\right)^2 f(r) u^0 \tag{8}$$

and

$$k^i \equiv 0 \tag{9}$$

Cold plasma is defined from [4] as plasma in which there exists a unique trajectory for an electron passing through specific points in spacetime.

The unperturbed trajectory X^μ is labeled as λ^0 and follows the wordlines of electrons in the unperturbed plasma $\vec{\lambda}$, where λ_i defines the 3-dimensional manifold of the unperturbed trajectories as

$$X^\mu \equiv x^\mu(\lambda^\mu) \tag{10}$$

Let A^μ be the vector potential, defined as

$$A'^\mu = a^\mu(X'^\mu)e^{i\phi(X'^\mu)+c.c.} \tag{11}$$

where $c.c.$ indicates the complex conjugate element.

The cold plasma is defined as

$$A^\mu \equiv 0 \tag{12}$$

and the Landau gauge is chosen for a^μ as

$$a_0 \equiv 0 \tag{13}$$

The WKB approximation is here studied as $\Phi > > 1$; different cases will be treated. The wave vector reduces to k^μ as

$$k^\mu \equiv \left(\omega, \vec{0}\right), \tag{14}$$

which simplifies to

$$k^\mu \equiv k^0 \equiv \frac{\partial \psi}{\partial x_\mu} \equiv \frac{\partial \psi}{\partial x_0} \tag{15}$$

Thus, in the case of cold plasma, the wave vector is defined based solely on the x_0 partial derivatives of the phase

$$k^0 \equiv -\frac{\partial \psi}{\partial x_0} \tag{16}$$

The Lagrangian density $\mathcal{L}_2$ is expressed as

$$\mathcal{L}_2 = \frac{1}{4\pi}\left[\omega_{pl} - k_\alpha g^{\alpha\beta} k_\beta a_\sigma a^{*\sigma} + \left|k_\mu a^\mu\right|^2\right] \tag{17}$$

For cold plasma, the Lagrangian density $\mathcal{L}_2$ from Eq. (17) is specified as

$$\mathcal{L}_{2\ cold\ \ plasma} \equiv \frac{1}{4\pi}\left[\omega_{pl}^2 - \left(k^0\right)^2 f(r) a_i a^{*i}\right] \tag{18}$$

The Lagrangian density $\mathcal{L}_{2\ cold\ \ plasma}$ is required to be stationary under variations of $a^{*\mu}$, leading to a definition of a generic Lagrange multiplier Ξ defined as

$$\Xi \equiv \left(k_i a^i\right)\left(k^\beta u^\beta\right). \tag{19}$$

The Lagrange multiplier Ξ in Eq. (19) vanishes identically for cold plasma

$$\Xi \equiv 0 \Leftrightarrow k_i a^i \equiv 0 \tag{20}$$

as $\vec{K} \equiv 0$ for cold plasma.

The effective propagation is identified as transverse waves

$$a_\mu k^\mu = 0 \tag{21}$$

Therefore, the Lagrangian density $\mathcal{L}_{2\ cold\ plasma}$ is specified as

$$\mathcal{L}_{2\ cold\ plasma} = \frac{1}{4\pi} F a_\mu a^{*\mu} \equiv F a_i a^{*i} \equiv F a_i g^{ij} a_j^* \tag{22}$$

where the $F = 0$ identically is studied.

Effective propagation is described with respect to two transverse waves $a_\mu a^\mu = 0$ and F_T as

$$F_T = \omega_{pl}^2 - k^\alpha k^\beta g_{\alpha\beta} = \omega_{pl}^2 - \left(k^0\right)^2 f(r). \tag{23}$$

The longitudinal wave is electrostatic in cold plasma, with F_L rewritten as

$$F_L \equiv \omega_{pl}^2 - k_\mu u^\mu = \omega_{pl}^2 - k_0 u^0 \tag{24}$$

The spectrum of quasi-bound states differs from that of the Proca equation in [5–7]. From [4], the quantum-gravitational Hamiltonian of a photon is calculated by ensuring the vanishing of the Hamiltonian constraint.

The dynamics are analyzed, particularly with respect to the Birkhoff Theorem, which is obeyed in Kottler-Schwarzschild-Kiselev spacetime. The group velocity v_g is calculated.

The Lagrangian S_2 is requested to be invariant under perturbations of the phase Φ. Its Lagrangian density $\mathcal{L}_2$ does not depend on Φ, but on its derivatives k_μ.

The following specifications in the case of cold plasma hold

$$a_i g^{ij} k_j = -a_i \frac{\partial \psi}{\partial x_i} = -a_i g^{ij} \frac{\partial \psi}{\partial x^j} \tag{25}$$

The mass of the photon in cold plasma is considered with its gravitational mass after the position

$$\xi_\mu U^\mu = 0 \equiv \xi^j g_{ij} u^j. \tag{26}$$

The Lagrangian density $\mathcal{L}_\omega$ is obtained after the continuity equation. The conserved current $J \equiv J(\omega, v_g)$ is given as $J(\omega, v_g) = v_g \mathcal{L}(\omega)$. Therefore, the density of the wave action is one evolving as the group velocity.

The Lagrangian density $\mathcal{L}_\omega$ is written as

$$\mathcal{L}_\omega = -\frac{1}{4\pi}\left[2k^0 a_\mu a^{+\mu} - \left(a_0 k_\mu a^{*\mu} + a^{*0} k_\mu a^\mu\right)\right] \tag{27}$$

and is specified in the case of cold plasma as

$$\mathcal{L}_\omega \equiv -\frac{1}{2\pi} k^0 a_i a^{*i} \tag{28}$$

The total number of photons N_{phot} is calculated as

$$N_{phot} = \int \sqrt{-g} d^3x \mathcal{L}_\omega : \tag{29}$$

It is required that the integrand vanishes at infinity for a finite wave packet. This hypothesis is valid, as the Kottler-Schwarzschild-Kiselev spacetime metric complies with the Birkhoff Theorem.

The number of photons remains conserved over time, verified by the vanishing time derivative.

$$\frac{d}{dt} N_{phot} = 0 \tag{30}$$

An analytical description of a wave packet propagating at a velocity v_g is investigated. The following analysis is deemed to be well-posed.

The averaged value of the mean position of the wave packet shell $< x^\mu >$ is calculated as

$$< x^\mu > = \frac{\int \sqrt{-g} d^3x x^\mu \mathcal{L}_\omega}{\int \sqrt{-g} d^3x \mathcal{L}_\omega} \tag{31}$$

The averaged value of the momentum of the wave packet shell $< x^\mu >$ is written as

$$< k_\mu > = \frac{\int \sqrt{-g} d^3x \left(\frac{\partial \psi}{\partial x^\mu}\right) \mathcal{L}_\omega}{\int \sqrt{-g} d^3x \mathcal{L}_\omega} \tag{32}$$

In the case of cold plasma, it is expressed as

$$< k_\mu > \equiv < k_0 > \equiv \frac{\int \sqrt{-g} d^3x \left(\frac{\partial \psi}{\partial x^0}\right) \mathcal{L}_\omega}{\int \sqrt{-g} d^3x \mathcal{L}_\omega} \tag{33}$$

Covariance is ensured by the vanishing of the covariant derivative of the Lagrangian density $\mathcal{L}_2$ as

$$\mathcal{L}_{2;\mu} = 0 \tag{34}$$

The variation of the function F with respect to ω is evaluated as

$$< \frac{\partial F}{\partial \omega} > = \frac{\int \sqrt{-g} d^3x \left(\frac{\partial F}{\partial \omega}\right) \mathcal{L}_\omega}{\int \sqrt{-g} d^3x \mathcal{L}_\omega} \tag{35}$$

The time evolution of the wave packet is calculated as

$$\frac{dx^\mu}{dt} = <(\partial F/\partial x_m u)_{x_{alpha}}/(\partial F/\partial \omega)> \tag{36}$$

The evolution of its 4-momentum is expressed as

$$\frac{dk^\mu}{dt} = <(\partial F/\partial x^m u)_{x_{alpha}}/(\partial F/\partial \omega)> \tag{37}$$

New features are identified, including the possibility of non-localized wave packets and the existence of multiple wave vectors.

4. The stress-energy tensor

The stress-energy tensor of the medium ${T^{\mu\nu}}_{medium}$ is specified in the case of cold plasma as

$${T^{\mu\nu}}_{medium} = n(m_e + m_p)U^\mu u^\nu \equiv {T^{00}}_{medium} \equiv {T^{00}}_{medium} = n(m_e + m_p)u^0 u^0 \tag{38}$$

The stress-energy tensor of the transverse waves ${T^{\mu\nu}}_{waveT}$ in the case of cold plasma is expressed as

$${T^{\mu\nu}}_{wave} = -\frac{a_\mu a^{*\mu}}{2\pi}\frac{\partial F}{\partial g_{\mu\nu}} \equiv -\frac{a_i a^{*i}}{2\pi} k^0 u^0 \tag{39}$$

The stress-energy tensors of both the medium and the waves obey the covariant derivative equality

$$T^{\mu\nu}{}_{;\mu} = \left({T^{\mu\nu}}_{medium} + {T^{\mu\nu}}_{wave}\right)_{;\mu} \equiv 0 \tag{40}$$

5. The massless limit of a Proca field on curved spacetime with plasma

As far as the second possibility is concerned, the massless limit of a Proca field on curved spacetime with plasma is considered in [8, 9] from [10]. The system is shown to admit plasma-driven quasi-bound electromagnetic states.

The results are specified here for the Kottler-Schwarzschild-Kiselev spacetime, where new terms in the system of differential equations are identified.

The case

$$lim_{\,r\to\infty} V \simeq \omega_{pl}^2 - \omega^2 \tag{41}$$

is studied, with ω being the photon frequency.

The 4-vector spherical harmonics are chosen as

$$Z_\mu^{(1)\ lm} = [1{,}0{,}0{,}0]Y^{lm} \tag{42}$$

$$Z_\mu^{(2)\ lm} = \left[0{,}f(r)^{-1}{,}0{,}0\right]Y^{lm} \tag{43}$$

$$Z_{\mu}^{(3)\ lm} = \frac{r}{\sqrt{l(l+1)}}\left[0, 0, \partial_{\theta}, \partial_{\phi}\right] Y^{lm} \tag{44}$$

$$Z_{\mu}^{(4)\ lm} = \frac{r}{\sqrt{l(l+1)}}\left[0, 0, \frac{\partial_{\phi}}{sin\theta}, \partial_{\theta}\right] Y^{lm} \tag{45}$$

These form an orthonormal basis as

$$\int dWZ_{\mu}^{(i)\ lm} \eta^{\mu\nu} Z_{\nu}^{(i')\ l'm'} = \delta^{ii'}\delta^{ll'}\delta^{mm'} \tag{46}$$

with $dW \equiv sin\theta d\theta d\phi$.

The representation in the frequency domain is developed as

$$h_{(i)}^{lm}(t, r) = h_{(i)}^{lm}(r) e^{-i\omega t} \tag{47}$$

The perturbations of the vector potential are written as

$$a_{\mu}(t, r, \theta, \phi) = \frac{1}{r}\sum_{i=1}^{i=4}\sum_{l,m} C_i h_{(i)}(t, r) Z_{\mu}^{(i)\ lm}(\theta, \phi) \tag{48}$$

with $C_1 = C_2 = 1$ and $C_3 = C_4 = 1/\sqrt{l(l+1)}$.

The field equations are rewritten, with the dependence on the parameters k and Λ qualifying the Kottler-Schwarzschild-Kiselev metric as

$$h_{(1)} = 0 \tag{49}$$

$$\left[r\left(l + l^2 + r^2\omega_{pl}^2\right)f - r^3\omega^2\right]h_{(2)} - r^2 f^2 h'_{(3)} = 0 \tag{50}$$

$$l(1+l)rfh_{(2)} + r^3\left(\omega^2 - f\omega_{pl}^2\right)h_{(3)} - l(l+1)r^2 f h'_{(2)} + 2Mrfh'_{(3)} + r^3 f^2(r) h''_{(3)} + \mathcal{O}(c_1; c_2) = 0 \tag{51}$$

$$\left[r\left(l + l^2 + r^2\omega_{pl}^2\right)f + r^3\omega^2\right]h_{(4)} - 2Mrfh'_{(4)} - r^3 f^2(r) h''_{(4)} + \mathcal{O}(c_1; c_2) = 0 \tag{52}$$

where the apices $'$ denote the classical partial derivative with respect to the radial direction, i.e. $h'_{(i)} \equiv \partial_r h_{(i)}/\partial r$.

The new terms found are those referring to $f(r)$ and one from Eq. (51) found as $\mathcal{O}(c_1)$. In particular, the term due to $\mathcal{O}(c_1; c_2)$ is of a different compared to those found from $f(r)$ in Eq. (2).

For the sake of investigation, it is useful to remark that Eqs. (49)-(52) hold also in the case of $\omega_{pl} \equiv \omega_{pl}(r)$.

The decomposition Eqs. (49)-(52) reveals the following.

The even-parity polar functions $h_{(1)}$ and $h_{(2)}$ are completely decoupled from the odd-parity function $h_{(4)}$, owing entirely to the spherical symmetry of the background metric.

For axial functions, the frequency ω_{pl} acquires the role of an effective mass for $h_{(4)}$.

For the polar general functions, key differences arise with respect to the Proca case as follows.

For $h_1 = 0$, one finds

$$\left(\omega^2 - \omega_{pl}^2\right)h_{(2)} = 0; \tag{53}$$

in the general case $l \neq 0$, the equations defining h_1 and h_2 are coupled within the Schrödinger-like equation

$$\frac{d^2}{dr^2}\Psi - V(r) = 0. \tag{54}$$

The behavior of the function with respect to the infinite-redshift surfaces of the Kottler-Schwarzschild-Kiselev metric is investigated.

The general limit is obtained from Eq. (54) to hold

$$lim_{\,r\to\infty}V(r) = -\omega^2 + lim_{\,r\to\infty}\omega_{pl}(r). \tag{55}$$

The Landau gauge is imposed, studying the limit

$$Z_0^{(i)\ \ lm} = 0 \tag{56}$$

From this, the orthonormal basis $\left\{\ Z_1^{(i)\ \ lm}, Z_2^{(i)\ \ lm}, Z_3^{(i)\ \ lm}\right\}$ is derived for the corresponding space.

6. Hot unmagnetized plasma

The Kottler-Schwarzschild-Kiselev metric is written as the line element

$$ds^2 = A(r)dt^2 + B(r)dr^2 + r^2(sin\theta)^2d\phi^2 + r^2d\theta^2 \tag{57}$$

with the specifications $A(r) \equiv f(r)$ and $B(r) \equiv 1/f(r)$, with $f(r)$ taken from Eq. (2).

The case of hot unmagnetized plasma is analyzed through the velocity

$$v_q^2 = \frac{\omega^2}{k_\mu k^\mu} \equiv \frac{k_0k^0 + \omega_p}{k_\mu k_\mu} \tag{58}$$

The velocity is computed using $u^\mu \equiv \left(u^0, \vec{u}\right)$ with $\vec{u} \equiv (u^r,0,0)$.

The elements of the new vector k^μ are written as

$$k^0 = p^0 - p_0u^0A(r)u^0 + p^rB(r)u^ru^0, \tag{59}$$

$$k^r = p^r - p^0u^0A(r)u^r + p^ru^rB(r)u^r. \tag{60}$$

The Lagrangian density $\mathcal{L}_\omega$ is calculated in the Landau gauge of the perturbation of the vector potential $a_0 = 0$ as

$$\mathcal{L}_\omega = -\frac{1}{41pi}\left[2k^0a_ia^{*i} - \left(-\frac{a^0K^ra^r}{f(r)} - a^{*0}\frac{K^ra^r}{f(r)}\right)\right] \tag{61}$$

From Eq. (17), the Lagrangian density $\mathcal{L}_2(K)$ is calculated from the Eqs. (59)-(60) as

$$\mathcal{L}_2(K) = \frac{1}{4\pi}\left[\omega_{pl} - \left[+(K_0)^2 f(r) - \frac{(K_r)^2}{f(r)}\right] a_\sigma a^{*\sigma} - \left|\frac{K_r a^r}{f(r)}\right|^2\right] \tag{62}$$

6.1 Large adiabatic perturbations of the plasma

Here, large adiabatic perturbations of the plasma are considered, with respect to those of dust and gas.

The Lagrangian density is divided into leading-order terms and perturbation orders, with specific focus on large adiabatic perturbations of the plasma. In particular, large adiabatic perturbations of the plasma are considered.

From the Lagrangian density $\mathcal{L}_2$, the following addends are found.

From

$$\mathcal{L}_2 = \frac{1}{4\pi}\left[\omega_{pl} - k_\alpha g^{\alpha\beta} k_\beta a_\sigma a^{*\sigma} + \left|k_\mu a^\mu\right|^2\right] \tag{63}$$

one obtains

$$p^0 = 0 \tag{64}$$

and the further constraints agree with those found in the study of the propagations of the waves.

The effective propagations described with respect to two transverse waves $a_\mu a^\mu = 0$ and F_T is specified as

$$F_T = \omega_{pl}^2 - k^\alpha k^\beta g_{\alpha\beta} = \omega_{pl}^2 + \left(k^0\right)^2 f(r) \tag{65}$$

with the constraint imposed after the perturbation of the velocity $\delta\vec{v}$.

The longitudinal wave is specified after F_L, rewritten as

$$F_L \equiv \omega_{pl}^2 - k_\mu u^\mu = \omega_{pl}^2 \tag{66}$$

with the constraint

$$K_r\delta(u^r) = 0 = (p^r - p^r\delta(u^r)B(r)\delta(u^r))\delta(u^r): \tag{67}$$

at each order, Eq. (67) is solved as

$$p^r \simeq 0. \tag{68}$$

The effects of the perturbation of the adiabatic perturbations of the velocities are therefore found.

As a result, only adiabatic perturbations of pressureless hot plasma are suitable for surrounding the Kottler-Schwarzschild-Kiselev black hole spacetimes.

The cosmological implementation for an expanding universe is discussed in [2]. Configurations of cold plasma and rarefied hot plasma are examined around black holes, including considerations for non-uniform plasmas.

The study is further refined for for adiabatic perturbations of the velocities, and representations of the accretion of the discs are presented. These plasma configurations are compared with those cosmologically allowed within the thermal history of the universe.

7. Prospective analysis of the dynamical magnetosphere

The dynamical magnetosphere of the Schwarzschild black hole spacetime is studied in [11], where the dispersion relations for cold plasma surrounding the Schwarzschild black hole, in the vicinity of the horizon, are analyzed [11]. Fourier analysis is applied to the magnetosphere in the background using the $3+1$ ADM splitting formalism, with the assumption of plane waves for the sake of simplification. Dispersion relations are further calculated in [12], under the same assumption of plane waves and Fourier decomposition.

The line element in the ADM formalism is written as

$$ds^2 = -\alpha^2 dt^2 + \gamma_{ij}\left[dx^i + \beta^i dt dx^j + \beta^j dx^i dt\right]. \tag{69}$$

The 4-velocity $\mathbf{n}$ in the fiducial configuration of the observer is perpendicular to the hypersurfaces of constant time, as from [13], as

$$\mathbf{n} = \frac{1}{\alpha}\left(\frac{\partial}{\partial t} - \beta^i \frac{\partial}{-} \partial x^i\right) \tag{70}$$

The Maxwell equations in the $3+1$ splitting from [14] are given as function of the perturbed magnetic field $\vec{B}$, perturbed velocity $\vec{V}$, and perturbed density ρ as

$$\vec{B} = \vec{B}^0 + B\vec{b}, \tag{71}$$

$$\vec{V} = \vec{V}^0 + \vec{v}, \tag{72}$$

$$\rho = \rho^0 + \rho\tilde{\rho}. \tag{73}$$

These equations depend on the variables z and t.

In the case of a non-rotating background, the General Relativistic Magnetohydrodynamics (GRMHD) equations are independent of the vanishing magnetic field.

The phase velocity v_p, associated with the propagation of the “induced signal” within the “carrier wave,” is calculated as

$$v_p = \omega/k. \tag{74}$$

The refractive index is defined as the inverse of the phase velocity.

The change rate of the refractive index with respect to the angular frequency $\frac{dn}{d\omega}$ allows one to analytically calculate the change rate of the inverse refractive index with respect to the phase velocity.

The group velocity v_g representing the features of the wave packet is calculated as

$$v_g = \frac{1}{n + \omega\frac{dn}{d\omega}}. \tag{75}$$

From [15], the macroscopic behavior of ultra-relativistic plasma shows that right and left circularly polarized waves propagate with identical dispersion relations. This differs from non-relativistic electron-ion plasma, where both the electron gas and positron gas are assumed to have the same temperature.

In the case of cold plasma in a rotating non-magnetized background, the perturbed variables taken into account are function of the variables t and $z$as

$$\tilde{\rho} \equiv \tilde{\rho}(t,z), \tag{76}$$

$$v_z \equiv v_z(t,z), \tag{77}$$

$$v_x = v_x(t,z). \tag{78}$$

In the case of rotating magnetized background, the magnetic field on the (x,z) plane is written according to the direction vectors $\hat{x}$ and $\hat{z}$ as

$$\vec{B} = B[\lambda(z)\hat{x} + \hat{z}] \tag{79}$$

with a velocity

$$V = \frac{V_F}{\alpha} + \lambda u \tag{80}$$

where V_F is the integration constant. The 3-dimensional mass-conservation constant allows the expression

$$u \equiv \frac{1}{1+z^2} \tag{81}$$

The quantity $\vec{b}$ is calculated as

$$\vec{b} = \frac{\delta\vec{B}}{|\vec{B}|} \tag{82}$$

whose components are $b_x(t,z)$ and $b_z(t,z)$.

Negative-phase velocities are also analyzed. It is observed that these negative-phase velocities propagate independently of the rotation of the black hole, enabling the formalism to be applied to rotating black hole cases.

8. Adiabatic perturbations of the accretion disc

From [16], the Lagrangian perturbation governing the adiabatic oscillation of thin, non-self-gravitating rotation discs is analyzed. The case considered here is one of the simplest, with $dM/dt \equiv 0$, where M represents the constant mass of the system.

The perturbation of the equation for inviscid fluid is written as

$$(\partial_t + v^\nu \nabla_\nu)v^\mu + \rho^{-1}\nabla^\mu P + \nabla^\mu \Phi = 0 \tag{83}$$

Here, Φ represents a (possibly non-Newtonian) potential.

8.1 The case of non-uniform plasma distributions

The outline from [17] is further developed here for non-uniform plasma distributions.

Given u, the phase velocity, and

$$n = \frac{c}{u} \tag{84}$$

the refractive index, the following properties of the refractive index are derived

$$n^2 = 1 + \frac{P_\mu P^\mu}{p_0 \sqrt{-g_{00}}} \tag{85}$$

Astatic medium is specified as

$$n^2 = 1 - \frac{\omega_{pl}}{\omega_{ph}}. \tag{86}$$

The effective photon mass is written as

$$m_{eff} = \frac{\hbar\omega(x^\alpha)}{c^2}\sqrt{1-n^2} = \frac{\hbar\omega^2}{c^2} \tag{87}$$

The corresponding effective velocity is given by

$$v_{eff} = \frac{|{}^3p|^2 c^2}{|{}^3p|^2 + m_{eff} p^2} : \tag{88}$$

Here, the effective velocity of the photon in plasma coincides with its group velocity.

8.2 Polytropic accretion

Polytropic accretion is studied following

$$\frac{p}{P_\infty} = \left(\frac{\rho}{\rho_\infty}\right)^\alpha \tag{89}$$

where P will be further studied in subsequent analyses as referenced in [18] and of [16].

From the refraction index, the following relation holds

$$n^2 = 1 + \frac{-|p^0|^2 + \left|\vec{p}\right|^2}{|p^0|^2} \tag{90}$$

from which

$$\left|\vec{p}\right|^2 = n^2 |p^0|^2 \tag{91}$$

and

$$p_\mu p^\mu = -m_{eff}c^2 = -\hbar\omega\sqrt{1-n^2}. \tag{92}$$

The group velocity V_{gr} is therefore defined as

$$V_{gr} = \left(\frac{\partial}{\partial\omega}(n\omega)\right)^{-1} c. \tag{93}$$

Given the Lorentz factor of the u velocity $\gamma(u)$, one finds

$$p^\mu = \gamma(u) m_0 u^\mu \tag{94}$$

such that

$$p_i p^i = \frac{m_0^2|u|^2}{1-\frac{|u|^2}{c^2}} \tag{95}$$

Using these tools, observational evidence from the "Project RadioAstron" should be verified.

The obtained results are further complemented by the analysis provided in [16].

8.3 Adiabatic accretion

The Lagrangian perturbation equations governing the adiabatic oscillation of thin, non-self-gravitating rotation discs are revisited from [16].

The dispersion relations that describe the oscillations within the disc are studied.

The trapping region is noted to depend on the structure of the unperturbed disc. A gas-pressure-dominated disc is used in this construction.

For the innermost stable orbit, the Lagrangian displacements are defined as

$$\xi_* = \xi(r,z)e^{i(m\phi+\sigma t)} \tag{96}$$

8.4 Cold plasma

The cold plasma configurations are newly analyzed within the framework of the limit

$$p^\mu \equiv lim_{\,^3p\to 0}(\rho, p) \equiv P^\mu \tag{97}$$

The effective mass m_{eff}^2 is calculated using the relation

$$p_\mu p^\mu = -m_{eff}^2 c^2 = \frac{\hbar\omega}{c^2}\sqrt{1-n^2} = P_0 P^0 \tag{98}$$

Derived from the refractive index.

The limit of the vanishing ^3p implies the identity

$$P_0 P^0 = \rho^2 f. \tag{99}$$

The constant K_3 is fixed after the identity

$$K_3 = u^0 f u^0 \tag{100}$$

which relates the quantities inherent to entropy.

The effective mass is calculated as

$$m_{eff}^2 = P^0 P_0. \tag{101}$$

8.5 Rarefied hot plasma

Here, the cases of rarefied plasmas are newly considered.

In the case of rarefied (non-interacting) plasma, the plasma frequency and the photon frequency are expressed as

$$\omega_{pl} < < \omega_{ph}, \tag{102}$$

The outline from [17] is further developed here. The frequency ω_{pl} is calculated after the dispersion relations obtained from the wave vectors, specified as

$$k^0 = p^0{}_{pl} f U^0 U^0, \tag{103}$$

$$k^r = p^0{}_{pl} f U^0 U^r. \tag{104}$$

Eq. (103) describes radio-frequency photons. Gravitational lensing, as described in [19, 20], is found to be proportional to

$$\hat{\alpha} = \frac{r_0}{b}\left[1 + \frac{1}{1 - \frac{\omega_{pl}^2}{\omega_{ph}^2}}\right]. \tag{105}$$

The interesting case where $\omega_{pl} = const$ is implied after the new position of u^r as

$$u^r \equiv const \tag{106}$$

From Eq. (106), the new constant K_5 is derived as

$$P^0{}_{pl} f u^0 = const = K_5 \tag{107}$$

As a further result, Eq. (107) K_5 fixes the values of the entropy.

8.6 Adiabatic perturbations

Adiabatic perturbations of the accretion disc are studied in [18], while new insights are derived from [16]. The non-relativistic theory of Lagrangian perturbations for adiabatic fluids is followed as outlined in [21]. In [18], the General-Relativistic effects on the epicyclic frequency of a gaseous disc around a Schwarzschild black hole are studied.

The General-Relativistic expression of ∇ is extended here without further modifications for the following cases.

The case $u^r = const$ is specifically considered.

To this end, the adiabatic perturbation is written from Eq. (2.1) [18] as

$$v^\nu \partial_\nu v^\mu + \nabla^\mu \Phi = 0 \equiv v^0 \nabla_0 v^\mu + \nabla^\mu \Phi = 0. \tag{108}$$

This is subsequently split due to

$$\nabla^\mu \Phi = \nabla^i \Phi. \tag{109}$$

In the case $u^r \equiv r^{-2+\varsigma}$, one still finds $\partial_t u^r = 0$ in multiple new configurations, such that the adiabatic perturbation is rewritten as

$$(\partial_t + v^\nu \nabla_\nu) v^\mu + \rho^{-1} \nabla^2 P + \nabla^\mu \Phi = 0. \tag{110}$$

Eq. (110) becomes

$$\left(v^0 \nabla_0 + v^r \nabla_r\right) v^\mu + \rho^{-1} \nabla^\mu P + \nabla^\mu \Phi = 0. \tag{111}$$

Eq. (111) is further split according to the requirements of Eq. (109), addressing new features of the gravitational potential. The $_{\theta\theta}$ and $_{\phi\phi}$components of the Einstein Field Equations (EFEs) generate new constraints on the Kottler-Schwarzschild-Kiselev potential, which are newly implemented here.

The case of adiabatic perturbations of stationary Newtonian fluid is described using Eq. (2.7) [18], descending from the following analysis

$$\Delta v^i = \partial \xi^i_*, \tag{112}$$

$$\delta \rho = -\rho \nabla_i \xi^i_*, \tag{113}$$

$$\frac{\partial P}{P} = \gamma \frac{\Delta \rho}{\rho} \tag{114}$$

The case of rarefied plasma is newly investigated for the instance $u^r = const$ as

$$\nabla v^i = \partial_t \xi^i_* = 0 \tag{115}$$

which implies

$$\partial_t \xi^i_* = 0 \tag{116}$$

It is determined that ξ^i_* does not depend on time within the present specification. Eq. (2.7) [18] describes the perturbation equations for stationary Newtonian fluid, where dependence is solely on the newly defined potential Φ. The new equations found here correspond again to perturbations of Newtonian fluid, in which the only General-Relativistic-derived quantity is the potential $\tilde{\phi}$.

The adiabatic perturbations of Newtonian fluid are written as

$$\partial_t^2\xi^\mu_* + 2\rho v^\nu \partial_t \xi^\mu_{*;\nu} + (1-\gamma)P^{,\mu}\xi^\nu_{*;\nu} - \gamma P \xi^{\nu,\mu}_{*;\nu} - P_{,\nu}\xi^{\nu;\mu}_{*;j} - \gamma^{,\mu} P \xi_{*;\nu} + \rho v^\nu v^\upsilon{}_{;\nu}\xi^\mu_{*;\upsilon} + \rho v^\nu v^\upsilon + \rho v^\nu v^\upsilon \xi^\mu_{*;\upsilon;\nu} + \rho\Phi^{,\mu}{}_{;\nu}\xi^\nu_* = 0 \tag{117}$$

where Φ represents the potential.

The new equations are implements for non-Newtonian potentials by splitting Eq. (116) as

$$(1-\gamma)P^{,\mu}\xi^\nu_{*;\nu} - \gamma P \xi^{\nu,\mu}_{*;\nu} - P_{,\nu}\xi^{\nu;\mu}_{*;j} - \gamma^{,\mu} P \xi_{*;\nu} + \rho v^\nu v^\upsilon{}_{;\nu}\xi^\mu_{*;\upsilon} + \rho v^\nu v^\upsilon + \rho v^\nu v^\upsilon \xi^\mu_{*;\upsilon;\nu} + \rho\Phi^{,\mu}{}_{;\nu}\xi^\nu_* = 0 \tag{118}$$

and

$$(1-\gamma)P^{,i}\xi^\nu_{*;\nu} - \gamma P \xi^{\nu,i}_{*;\nu} - P_{,\nu}\xi^{\nu;i}_{*;\nu} - \gamma^{,i} P \xi_{*;\nu} + \rho v^\nu v^\upsilon{}_{;\nu}\xi^i_{*;\upsilon} + \rho v^\nu v^\upsilon + \rho v^\nu v^\upsilon \xi^i_{*;\upsilon;\nu} + \rho\Phi^{,i}{}_{;\nu}\xi^\nu_* = 0 \tag{119}$$

Furthermore, the saturation of $P_{,\nu}\xi^{\nu;i}_{*;\nu}$ in Eq. (119) is required to be further divided.

9. Comments

The study of the non-Newtonian potential involves comparing the Novak-Wagoner potential Φ_{NW}

$$\Phi_{NW} = -\frac{GM}{r}\left[1 - 3\frac{GM}{rc^2} + 12\frac{G^2M^2}{r^2c^4}\right]. \tag{120}$$

with the Muchotrzeb-Paczynski potential [22] and the gravitational potential of the Kottler-Schwarzschild-Kiselev metric Φ_{KSK} [23].

In [22], the structure of the accretion disc near the innermost stable orbit is analyzed. The potential takes into account the gradient of pressure and dynamical terms. Non-Keplerian effects, such as radial heat diffusion, are also included. A critical rate dM/dT is defined, describing the transition from subsonic radial flow to supersonic radial flow near the inner disc radius.

In [18], the General-Relativistic effects are studied in regions between $3r_S$ and $4r_S$, where trapped oscillations are observed. The analysis considers black holes with masses $10^9 M_\odot - 10^{10} M_\odot$.

The potential considered in [22], V_{MP}, is defined based on the acceleration due to radial velocity v_R as

$$\frac{d}{dr}v_R = \frac{1}{v_R\left(1 - \frac{a_s}{v_R^2}\right)}\left[\frac{a_s^2}{r} + a_s^2\frac{d}{dr}lnz + \left(\Omega^2 - \Omega_k^2\right)r\right] \tag{121}$$

where a_s is the sound speed of the flow.

From [23], the terms relevant for inclusion in the non-Newtonian potential, within the context of analyzing disc structures from $-\frac{d}{dr}\Gamma^r_{00}$ which are to be compared with the other two potentials Eqs. (120) and (121)

$$\Phi_{KSK} \simeq -\frac{1}{2}\left[\frac{r_S}{r} - c_1 - 2\frac{r_S^2}{r^3} - 2c_2r + 2c_1^2r + 6c_1c_2r^2\right] \tag{122}$$

The study for a non-vanishing dM/dt is compared with findings from [24].

9.1 Comparison with other theories of gravity

The comparison of the roles of the parameters c_1 and c_2 is conducted in the context of other theories of gravity, specifically the 4-dimensional conformal Weyl gravity [25, 26], and the correspondence between black hole solutions in conformal gravity and massive gravity [27, 28]. The well-posedness of the limit is examined here according to the curvature.

From Eq. (6) of [29].

$$q(r) = s_0 + \frac{s_2}{r} + s_1r + s_3r^2 \tag{123}$$

The comparison of black hole solutions between conformal gravity and massive gravity is expressed as

$$\tilde{q}(r) = \mathcal{K} + m^2C^2c_2 - \frac{M}{r} + m^2\frac{Cc_1}{2}r + s_3r^2 \tag{124}$$

$$s_0 = \mathcal{K} + m^2C^2c_2 \equiv 1, \tag{125}$$

$$s_1 = \frac{m^2Cc_1}{2}, \tag{126}$$

$$s_2 = -M. \tag{127}$$

A new constraint is posed as

$$s_0 = s_0^2 = 1 \tag{128}$$

or equivalently as

$$s_0 = 1 = \mathcal{K} + m^2C^2c_2 = \mathcal{K}^2 - 3s_1M = \mathcal{K}^2 - \frac{3}{2}Mm^2Cc_1 \tag{129}$$

After solving for M, the limit $m \to 0$ is found to be well-posed in both the asymptotically flat spacetime configuration $\mathcal{K} \to 0$, and in the case $\mathcal{K} = 1$.

The comparison is validated due to the adherence of the Kottler-Schwarzschild-Kiselev spacetimes to the Birkhoff Theorem, as discussed in [23].

9.2 Experimental verification: Emissions

Radio emissions are analyzed following the implementation of Eqs. (103) and (104).

The adiabatic variation in the gravitational potential of the self-gravitating accretion object is studied to understand the emissions, as from [30–32]. This framework also facilitates the interpretation X-ray emissions.

9.3 Comparison with the innermost stable circular orbits of Kerr-Newmann black hole spacetimes and removal of the degeneracies before evaporation

From [1], the role of the geometric parameter c_2 is studied as a substitute for the spin parameter of a Kerr black hole when accounting for the size of accretion objects (such as discs, analyzed therein [1]).

From [23], the Schwarzschild radius is expressed as

$$r_S = -\frac{1}{27}\frac{\left(2c_1^2 + 3c_2\right)^{3/2} + 2c_1^3 + 9c_1c_2}{c_2^2}. \tag{130}$$

Unlike the findings in [1], two specific determinations of the cosmological-constant parameter are newly derived and written explicitly as:

$$c_2 = \frac{1}{54}\left[-18c_1r_S + 1 \pm \sqrt{16c1^4r_S^2 - 432c_1^3r_S^3 + 216c_1^2r_S^2 - 36c_1r_S + 1}\right] \tag{131}$$

Also distinct from [1], the degeneracy is resolved in the non-Schwarzschild case $c_1 \neq 0, \quad c_2 \neq 0$, as the Birkhoff Theorem is demonstrated to hold in [23]. This allows for a direct comparison with the innermost stable circular orbits (ISCOs) of Kerr-Newman black hole spacetimes.As a result, the cosmological-constant parameter mimics the role of the spin parameter in determining the ISCOs of Kerr-Newman black hole spacetimes. This conclusion is reached by directly solving Eqs. (34) and (37) from [1].

The equations defining the ISCOs for Kerr-Newmann black hole spacetimes are found by solving Eq. (37) for r_{ISCO} from [1]. The definition of the ISCO is incorporated to remove degeneracy prior to black hole evaporation within its evolutionary timeline. For a black hole filled with matter, the ISCO is determined as the smallest value among the four possible solutions derived.

9.3.1 A new degeneracy

Furthermore, a new degeneracy is identified here concerning the parameter c_1, as noted in comparison with [1]. The solution of Eq. (130) with respect to c_1 brings four new possible expressions, which must to be validated within the parameter space of the models, such as in [23].

9.4 Experimental validation: The photon region and deflection of light in the strong-field limit

The deflection of light in the strong-field limit of the Kottler-Schwarzschild-Kiselev spacetimes is calculated according to the indications provided in [33]. In this context, the results from [33] are proven to hold, as the considered spacetimes obey the Birkhoff Theorem [23].

The radius of the photon sphere r_{PS} is newly derived as

$$r_{PS} = \frac{1}{3}\frac{3^{1/3}\left[3\left(\sqrt{3}\sqrt{1+27c_2r_s^2} + 9\sqrt{c2}r_S\right]^{2/3-3^{1/3}}}{\sqrt{c_2} - \left[\sqrt{3}\sqrt{1+27c_2r_s^2} + 9\sqrt{c2}r_S\right]^{1/3}} \tag{132}$$

The radius of the photon sphere as outlined in Eq. (132) depends solely on the cosmological constant parameter. This dependence highlights a role comparable to that of the spin parameter in determining the ISCOs of Kerr-Newman spacetimes.The deflection angle α_{defl} at x_0 is calculated as

$$\alpha_{defl} = -\pi + \int_{x_0}^{\infty} dr 2 \frac{1}{\sqrt{f(r)}} \frac{1}{r} \frac{1}{\sqrt{\frac{r^2}{r_{x_0}^2} \frac{f(r_{x_0})}{f(r)} - 1}} \tag{133}$$

The deflection angle in the strong-field limit is found to depend on both the parameter c_1 and c_2. A detailed comparison with the ISCOs of Kerr-Newman black hole spacetimes is required to further elucidate this relationship.

10. Conclusions

In this chapter, we thoroughly analyzed the Kottler-Schwarzschild-Kiselev black hole spacetimes. The propagation of photons was studied both in vacuum and in the presence of media. Specifically, the interactions between photons and cold unmagnetized plasma, as well as hot rarefied unmagnetized plasma, were detailed extensively.

The procedures taken from [34] allow one to define the complete history of the Kottler-Schwarzschild-Kiselev spacetimes surrounded of media, from the quantum-gravitational interactions of waves to the formation of accretion objects until evaporation.

The effects described in [35] of the Schwarzschild black hole horizon on isothermal plasma surfaces were reconciled with the configurations studied in [34] as particular cases. These new findings open the way for comparisons with spinning black hole objects, as referenced in [36, 37], specifically concerning the quantum-gravitational interactions of photons in the media surrounding black hole objects. The paradigms are validated against cosmological evolution scenarios from [2].

Unlike the scenarios proposed in [1] and the references therein, the degeneracies attributed to the presence of the cosmological-constant term are resolved. In addition, the comparison with the Schwarzschild spacetime is newly delineated according to the model's compliance with the Birkhoff Theorem [23]. A new degeneracy is also identified regarding the role of the linear term.

All elements necessary for experimental validation are compiled, including emissions from plasma jets, the photon region, and the deflection of light in the strong-field limit.

The paradigm is therefore ready for confrontation with other theories of gravity, such as 4-dimensional conformal Weyl gravity and the correspondence between black hole solutions in conformal gravity and massive gravity. Moreover, the geometrical role of the linear term and the cosmological constant term can be compared with the spin parameter in the Kerr-Newman black hole spacetime.

Conflict of interest

The author declares no conflict of interest.

Author details

Orchidea Maria Lecian
Sapienza University of Rome, Rome, Italy

*Address all correspondence to: orchideamaria.lecian@uniroma1.it

References

[1] Gregoris D, Ong YC, Wang B. A critical assessment of black hole solutions with a linear term in their redshift function. European Physical Journal C: Particles and Fields. 2021;**81**:684-690

[2] Chibisov GV. Damping of adiabatic perturbations in an expanding universe. Soviet Astronomy. 1972;**16**:56

[3] Kiselev VV. Quintessence and black holes. Classical and Quantum Gravity. 2003;**20**(6):1187-1198

[4] Kulsrud R, Loeb A. Dynamics and gravitational interaction of waves in nonuniform media. Physical Review D. 1992;**45**:525

[5] Pani P, Cardoso V, Gualtieri L, et al. Black hole bombs and photon mass bounds. Physical Review Letters. 2012; **109**:131102

[6] Pani P, Cardoso V, Gualtieri L, et al. Perturbations of slowly rotating black holes: Massive vector fields in the Kerrr metric. Physical Review D. 2012;**86**: 104017

[7] Brito R, Cardoso V, Pani P. Superradiance: New Frontiers in Black Hole Physics. Vol. 906. Springer; 2015

[8] Cannizzaro E, Caputo A, Sberna L, et al. Plasma-photon interaction incurved spacetime: Formalism and quasibound states around nonspinningblack holes. Physical Review D. 2021;**103**:124018

[9] Lecian OM. Dynamics and Quantum Gravitational Interactions of Waves in Kottler-Schwarzschild-Kiselev Space-Times Surrounded of Plasma and Comparison with Experimental Evidence. Berlin, Germany: Research Gate; 2024. DOI: 10.13140/ RG.2.2.34976.39681

[10] Rosa JG, Dolan SR. Massive vector fields on the Schwarzschild spacetime: Quasi-normal modes and bound states. Physical Review D. 2012;**85**:044043

[11] Sharif M, Sheikh U. Cold plasma dispersion relations in the vicinity of a Schwarzschild black hole horizon. General Relativity and Gravitation. 2007;**39**:1437-1465

[12] Wald RM. Black hole in a uniform magnetic field. Physical Review D. 1974; **10**:1680

[13] Zhang X-H. Interactions of magnetohydrodynamic waves with gravitomagnetic fields, and their possible roles in black-hole magnetospheres. Physical Review D. 1989;**40**:3858

[14] Thorne KS, Price RH, Macdonald DA. Black Holes: The Membrane Paradigm. New Haven: Yale University Press; 1986

[15] Ji S, Kawata T. Waves in an ultra-relativistic electron-positron plasma. Journal of the Physical Society of Japan. 1980;**49**:747

[16] Nowak M, Wagoner RV. Diskoseismology: Probing accretion disks. I. Trapped adiabatic oscillations. Astrophysical Journal. 1991;**378**:656

[17] Bisnovatyi-Kogan GS, Tsupko O. Yu: Gravitational lensing in a non-uniform plasma. Monthly Notices of the Royal Astronomical Society. 2010;**404**:1790

[18] Kato S, Fukue J. Publications of the Astronomical Society of Japan. 1980;**32**: 377

[19] Schneider P, Ehlers J, Falco E. Gravitational Lensing. Berlin: Springer-Verlag; 1992

[20] Schneider P, Kochanek CS, Wambsganss J. Gravitational lensing: Strong, weak and micro. In: Swiss Society for Astrophysics and Astronomy Series: Saas-Fee Advanced Courses, Number 33. Berlin: Springer; 2006

[21] Friedman JL, Schutz BF. Lagrangian perturbation theory of nonrelativistic fluids. Astrophysical Journal, Part 1. 1978;**221**:937

[22] Muchotrzeb B. Paczyński: Transonic accretion flow in a thin disk around a black hole. Acta Astronomica B. 1982;**32**:1

[23] Lecian OM. Generalized Schwarzschild spacetimes with a linear term and a cosmological constant. Universe. 2024;**10**(11):408

[24] Shakura NI, Sunyaev A. Black holes in binary systems. Observational Appearance. Astronomy and Astrophysics. 1973;**24**:337

[25] Klemm D. Topological black holes in Weyl conformal gravity. Classical and Quantum Gravity. 1998;**15**:3195

[26] Mannheim PD, Kazanas D. Solutions to the Reissner-Nordström, Kerr, and Kerr-Newman problems in fourth-order conformal Weyl gravity. Physical Review D. 1991;**44**:417

[27] Cai R-G, Hu Y-P, Pan Q-Y, Zhang Y-L. Thermodynamics of black holes in massive gravity. Physical Review D. 2015;**91**:024032

[28] Hendi SH, Eslam Panah B, Panahiyan S. Einstein-born-Infeld-massive gravity: adS-black hole solutions and their thermodynamical properties. Journal of High Energy Physics. 2015; **11**:157

[29] Eslam Panah B, Hendi SH. Black hole solutions correspondence between conformal and massive theories of gravity. EPL. 2019;**125**:60006

[30] Lecian OM. Transient phenomena of inviscid accretion gas-radiation slim disc in a gravitational potential after adiabatic perturbations of the velocities. International Journal of Mathematics and Computer Research. 2024;**12**:4562

[31] Vasyliunas VM. Theories of magnetospheres around accreting compact objects. Space Science Reviews. 1979;**24**(4):609-634

[32] Zhang XH. 3+1 formulation of general-relativistic perfect magnetohydrodynamics. Physical Review D. 1989;**39**(10):2933

[33] Bozza V. Gravitational lensing in the strong field limit. Physical Review D. 2002;**66**:103001

[34] Lecian OM. Non-rotating Spherically-Symmetric Blackholes. Chisinau: OmniScriptum S.R.L. LAP Lambert Academic Publishing; 2024

[35] Sharif M, Sheikh U. Effects of Schwarzschild black hole horizon on isothermal plasma wave dispersion. General Relativity and Gravitation. 2007;**39**:2095-2124

[36] Liu C-Q, Ding C-K, Jing J-L. Effects of homogeneous plasma on strong gravitational lensing of Kerr black holes. Chinese Physics Letters. 2017;**34**:090401

[37] Kimpson T, Kinwah W, Zane S. Spatial dispersion of light rays propagating through a plasma in Kerr space–time. Monthly Notices of the Royal Astronomical Society. 2019;**484**(2):2411-2419. DOI: 10.1093/mnras/stz138

Chapter 3

Perspective Chapter: Gravity in Extradimensional Space

Eugênio Maciel

Abstract

Since its inception as science, physics has aimed to describe the functioning of the Universe as a whole, from its smallest structures, such as elementary particles, to its largest ones, like galaxies and galaxy clusters. Through a unified framework, the four fundamental interactions should be fully integrated, stemming from a single source or field. This theory is known as the Grand Unification Theory (GUT). Apparently, these theories require additional spatial dimensions in our Universe for their formulation, even though current phenomenology does not yet support the existence of such extra dimensions. In this chapter, we aim to outline the main theories of extra dimensions, from the pioneering Kaluza-Klein theory to the so-called large-scale theories, such as the Randall-Sundrum models and the ADD model, which involve the concept of brane or membrane.

Keywords: gravity, extra dimensions, braneworld, Kaluza-Klein theory, large extra dimensions

1. Introduction

In 1916, Albert Einstein introduced to the world his General Relativity Theory (GRT), which forever changed our understanding of the Universe. Complete and self-consistent, this theory provided a new interpretation of the phenomenon of gravity, which had previously been described by Newton's Law of Universal Gravitation. The opposite of Newtonian thought, which considers gravity to arise from a gravitational field created by the mass of a massive body, the (GRT) regards gravity as a geometric property of spacetime, a quantity already defined in his Special Theory of Relativity.

After formulating the GRT, Einstein sought to develop a Unified Field Theory (UFT), a theory that would describe the two fields known at the time—the gravitational field and the electromagnetic field—through a single general theory. However, Einstein did not succeed in this endeavor. It was in the year 1921 that mathematician Theodor Kaluza published a brief article describing the unification of gravity and electromagnetism by considering a five-dimensional spacetime [1]. Five years later, in 1926, physicist Oscar Klein introduced some modifications to Kaluza's theory, attempting to provide a physical justification for it [2]. Since then, the theory has been known as the Kaluza-Klein theory. This was the first theory involving extra dimensions.

Over time, interest in the study of extra dimensions diminished somewhat, given the impossibility of empirical verification. It was only in 1998 that the idea resurfaced with the brane-world model, which aimed to solve the hierarchy problem—the significant energy difference between the Planck scale 10^{18} GeV and the electroweak scale 10^3 GeV. This model is known as the ADD model and represents the first large-scale extra-dimensional model [3].

There are two important large-scale extra-dimensional models: the Randall-Sundrum models, which differ essentially only in how the extra dimensions are compactified [4, 5].

2. The Kaluza-Klein theory

The unification of Maxwell's Electromagnetism and Einstein's General Relativity becomes possible when spacetime is considered to have five dimensions: one temporal dimension and four spatial dimensions. The fifth dimension, z, together with the four dimensions of ordinary spacetime, forms a $(4+1)-$ dimensional coordinate set described as (x^μ, z), where $\mu = 0,1,2,3$.

In Kaluza-Klein theory, the metric—an essential element for Metric Theories of Gravitation (e.g., General Relativity)—when observed in the physical spacetime, consists of three distinct parts:

- The metric of ordinary spacetime $g_{\mu\nu}$, where $\mu = 0,1,2,3$.
- A four-vector field $g_{\mu z} = g_{z\nu}$.
- A scalar field g_{zz}.

The component $g_{\mu z}$ can be identified as A_μ, the electromagnetic vector potential, and g_{zz} corresponds to the scalar field ϕ. This leads to a metric g_{AB}, parameterized as follows:

$$g_{AB} = \begin{pmatrix} g_{\mu\nu} + \kappa\phi A_\mu A_\nu & \kappa\phi A_\mu \\ \kappa\phi A_\nu & \phi \end{pmatrix} \tag{1}$$

the four $-$ dimensional metric is considered with a signature of the type $(+,-,-,-)$, and the indices of g_{AB} vary as $(A, B = 0,1,2,3,4)$.

As a starting point, Kaluza proposed that the Universe in five dimensions is free of matter. Under this condition, the General Relativity equations (Einstein equations) are described as follows:

$$G_{AB} = 0, \tag{2}$$

which is equivalent to

$$R_{AB} = 0, \tag{3}$$

where $G_{AB} = R_{AB} - \frac{1}{2} R g_{AB}$ is the five $-$ dimensional Einstein tensor. The quantity R_{AB} is defined as follows:

$$R_{AB} = \partial_C \Gamma^C_{AB} - \partial_B \Gamma^C_{AC} + \Gamma^C_{CD}\Gamma^D_{AB} - \Gamma^C_{BD}\Gamma^D_{AC}, \tag{4}$$

which is the Ricci tensor, and $R = g^{AB} R_{AB}$ is the scalar curvature.

The five – dimensional Ricci tensor retains the same form as in the ordinary case, and the Christoffel symbols, defined in terms of the metric g_{AB}, are given similarly.

2.1 Cylindrical condition

Phenomenologically, the extra dimension is not observed. To account for this, Kaluza imposed the cylinder condition in his model, requiring that the derivative of all metric components with respect to the extra coordinate vanish:

$$\frac{\partial g_{AB}}{\partial z} = 0. \tag{5}$$

Using this condition, we can determine the field equations in the context of extra dimensions. Based on Eqs. (2) and (5), and the metric g_{AB}, we obtain

$$G_{\mu\nu} = \frac{\kappa^2\phi^2}{2}T^{EM}_{\mu\nu} - \frac{1}{\phi}\left[\nabla_\mu(\partial_\nu\phi) - g_{\mu\nu}\Box\phi\right], \tag{6}$$

$$\nabla^\mu F_{\mu\nu} = -3\frac{\partial^\nu\phi}{\phi}, \quad \Box\phi = \frac{\kappa^2\phi^3}{4}F_{\mu\nu}F^{\mu\nu}. \tag{7}$$

where $\Box = \frac{\partial^2}{\partial t^2} - \frac{\partial^2}{\partial x^2}$ is the d'Alembert operator. Here, $G_{\mu\nu} = R_{\mu\nu} - \frac{1}{2}Rg_{\mu\nu}$ is the ordinary four-dimensional Einstein tensor. The tensor $T^{EM}_{\mu\nu} = \frac{1}{4}g_{\mu\nu}F_{\alpha\beta}F^{\alpha\beta} - F^\alpha_\mu F_{\nu\alpha}$ represents the electromagnetic energy-momentum tensor, and $F_{\alpha\beta} = \partial_\alpha A_\beta - \partial_\beta A_\alpha$ is the electromagnetic field tensor. If the scalar field ϕ is constant in spacetime, the Eqs. (6) and (7) reduce to the standard Einstein and source-free Maxwell equations:

$$G_{\mu\nu} \quad = 8\pi G\phi^2 T^{EM}_{\mu\nu}, \nabla^\mu F_{\mu\nu} \quad = 0. \tag{8}$$

From (2.1.9), the additional condition $F_{\mu\nu}F^{\mu\nu} = 0$ is satisfied for radiation fields.

2.1.1 Variational method

Most physical systems derive their equations of motion from Hamilton's principle, also known as the "principle of least action." According to this principle, the "action" function of the dynamic variables remains stationary under small variations of these variables. This method is advantageous because it establishes a connection between symmetries and conservation laws. In General Relativity, Einstein's equations are derived from the Einstein-Hilbert action:

$$S_{EH} = -\frac{1}{16\pi G}\int d^4x\sqrt{-g}R, \tag{9}$$

where $g = \det\left(g_{\mu\nu}\right)$ is the determinant of the four-dimensional metric and R is the Ricci scalar.

The Maxwell equations in the absence of sources are derived from the action:

$$S_{EM} = -\frac{1}{4}\int d^4x\sqrt{-g}F_{\mu\nu}F^{\mu\nu}. \tag{10}$$

In five dimensions, the Einstein-Hilbert action is expressed as follows:

$$S = \int d^5x \sqrt{\hat{g}} \hat{\mathfrak{R}}, \tag{11}$$

where $\hat{g}$ is the determinant of the metric $(-, +, \ldots, +)$ and $\hat{\mathfrak{R}}$ is the Ricci scalar in the higher-dimensional space. Using the metric decomposition and the cylinder condition, it can be shown that

$$S = -\int d^4x \sqrt{-g}\phi \left(\frac{R}{16\pi G} + \frac{1}{4}\phi^2 F_{\mu\nu}F^{\mu\nu} + \frac{2}{3\kappa^2}\frac{\partial^\mu \phi \partial_\mu \phi}{\phi^2} \right), \tag{12}$$

This reveals that General Relativity, when interpreted as a five-dimensional vacuum theory, includes General Relativity in four-dimensional spacetime with the addition of an electromagnetic field.

2.2 Klein compactification

In 1926, mathematician Oscar Klein aimed to provide a physical basis for Kaluza's cylinder condition. He proposed that the extra dimension should have the topology of a circle with a very small length scale. This compactification explains the seemingly four-dimensional nature of the Universe and became known as Klein's compactification mechanism, which, together with Kaluza's work, constitutes the Kaluza-Klein theory. **Figure 1** illustrates this idea simply.

The four spatial dimensions form a "cylinder" where the three usual dimensions (x^1, x^2, x^3) are infinite with the topology of $\mathbb{R}^3$. The extra dimension z is a circle of radius l, with topology S^1. Considering a scalar field $\phi(x^\mu, z)$ in a five-dimensional space, the compact topology requires periodicity along the extra dimension:

$$\phi(x^\mu, z) = \phi(x^\mu, z + 2\pi l), \tag{13}$$

where l is the radius of the fifth dimension. Due to this periodicity, the fields in the metric can be expanded as Fourier series:

$$g_{\mu\nu}(x^\mu, z) = \sum_n g_{\mu\nu}^{(n)}(x^\mu) e^{inz/l}, \tag{14}$$

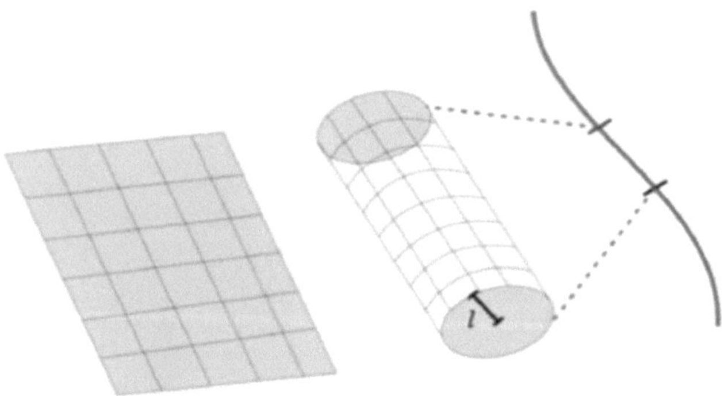

Figure 1.
The Kaluza-Klein Theory.

$$\phi(x^\mu, z) = \sum_n \phi^{(n)}(x^\mu) e^{inz/l}, \tag{15}$$

$$A_\mu(x^\mu, z) = \sum_n A_\mu^{(n)}(x^\mu) e^{inz/l}, \tag{16}$$

where n is the Fourier mode index. Fields are independent of the extra coordinate only for $n = 0$, corresponding to the zero mode.

For a scalar field in flat spacetime, the Klein-Gordon equation in five dimensions is

$$\Box_{(5)} \phi = 0, \tag{17}$$

where $\Box_{(5)} = \Box + \frac{\partial^2}{\partial z^2}$ is the five-dimensional d'Alembert operator. Using the method of separation of variables, we consider a solution of the form:

$$\phi(x^\mu, z) = \chi(x^\mu)\varphi(z), \tag{18}$$

where $\chi(x^\mu)$ depends only on x^μ and $\varphi(z)$ only on z. Substituting this into the Klein-Gordon equation gives

$$\frac{1}{\chi(x^\mu)} \Box \chi(x^\mu) + \frac{1}{\varphi(z)} \frac{\partial^2 \varphi(z)}{\partial z^2} = 0. \tag{19}$$

This equation is satisfied if

$$\frac{1}{\chi(x^\mu)} \Box \chi(x^\mu) = C, \ \frac{1}{\varphi(z)} \frac{\partial^2 \varphi(z)}{\partial z^2} = -C, \tag{20}$$

where C is a constant. The equation for $\chi(x^\mu)$ becomes

$$\Box \chi(x^\mu) = C\chi(x^\mu), \tag{21}$$

which is the Klein-Gordon equation in four-dimensional spacetime, where C represents the field's mass. The equation for $\varphi(z)$ is

$$\frac{\partial^2 \varphi(z)}{\partial z^2} = -C\varphi(z), \tag{22}$$

with the solution

$$\varphi(z) = A \sin\left(\sqrt{C} z\right) + B \cos\left(\sqrt{C} z\right). \tag{23}$$

This solution reflects the compact nature of the extra dimension. Thus, with the boundary conditions given by (13), we must have

$$\phi(0) = \phi(2\pi l). \tag{24}$$

From this, we observe that C cannot be arbitrary. The possible allowed values for C are expressed as

$$C = \frac{n^2}{l^2}. \tag{25}$$

We will see later that this result leads to an important conclusion. Noting that C is positive, we can rewrite it as $C = m^2$. For each allowed value of m, we must have

$$\Box\chi^{(n)}(x^\mu) = m^2_{(n)}\chi^{(n)}(x^\mu), \tag{26}$$

where

$$m_n^2 = \frac{n^2}{l^2}. \tag{27}$$

From this, we can see that, from a four-dimensional perspective, $\chi(x, z)$ can be decomposed into the following:

- A zero mode $n = 0$ that is massless $\phi^{(0)}$.
- A set of other modes, called Kaluza-Klein modes or simply (KK) modes.

In this case, each (KK) mode can be interpreted as a particle with mass $|n|/l$.

If l is sufficiently small, for instance, on the order of the Planck length $l \sim 10^{-35}$ m, as Klein suggested, the energy required to excite the nonzero modes $n \neq 0$ would be beyond the reach of current experiments. Thus, for the extra dimension to be observable, particle accelerators would need to achieve energy levels on the order of $E \sim 1/l$. This would explain why the extra dimension has not yet been detected.

3. Extra dimensions in large scale: The ADD model

We saw that, in Kaluza-Klein theory, the extra dimension is compact. If its length is proportional to the Planck length, it would be undetectable, as discussed in the previous section. Nearly 70 years later, physicists Nima Arkani-Hamed, Savas Dimopoulos, and Georgi Dvali proposed a model for extra dimensions that diverged in a crucial aspect from the Kaluza-Klein theory. In this model, known as the ADD model [3], the extra dimension is still compact, but its length can reach a submillimeter scale without conflicting with experimental observations. One of the motivations for this model is an attempt to address the well-known hierarchy problem. This problem, briefly summarized, refers to the significant energy disparity between the electroweak scale and the Planck scale.

The ADD model is based on the hypothesis that matter and fields are confined to a three-dimensional spatial space (a three-brane), while gravity can propagate along the extra dimension. In this section, we will study the localization of fermions on the brane [6] and also discuss the effects of the extra dimension about gravity in this model.

3.1 Confinement of matter

We can imagine our Universe in $(3 + 1)$ dimensions as a hypersurface embedded in a space with extra dimensions. According to the brane model, our ordinary hypersurface is confined and evolves within a higher dimensional space. The spatial part of this hypersurface is called the three-brane, where the number "3" refers to the number of spatial dimensions. Therefore, a brane is a submanifold of a larger ambient space.

A simple way to illustrate the mechanism of matter confinement is to consider a theory of fermion localization by a domain wall in a model that assumes the existence of the extra dimension z [7]. In this case, the domain wall is described by a scalar field $\phi = \phi(x^\mu, z)$. The action associated with the scalar field is given by

$$S = \int d^4x, dz \left[\frac{1}{2} (\partial_A \phi)^2 - V(\phi) \right], \tag{28}$$

where $A = 0,1,2,3,4$ and $V(\phi)$ is the scalar potential. Let us consider the potential $V(\phi)$:

$$V(\phi) = \frac{\lambda^2}{8} \left(\phi^2 - \nu^2 \right)^2. \tag{29}$$

It is easy to observe that $V(\phi)$ has two stable energy minima, at $\phi = -\nu$ and $\phi = \nu$, and an unstable maximum at $\phi = 0$. We observe this behavior in **Figure 2**.

Our goal is to find a solution of the domain wall type. To achieve this, we will look for a static solution that depends only on the extra dimension z. We must take into account the dynamics of the field, which is obtained by applying the Euler-Lagrange equation $\frac{\partial \mathscr{L}}{\partial \phi} - \partial_A \left[\frac{\partial \mathscr{L}}{\partial(\partial_A \phi)} \right] = 0$ to the Lagrangian shown in (28). This yields the following result:

$$\Box_A \phi + \frac{dV}{d\phi} = 0. \tag{30}$$

Here, $\Box_A^{(5)} = \partial_A^2 = \eta^{AB} \partial_A \partial_B$, and in this case,

$$-\frac{d^2 \phi_0(z)}{dz^2} + \frac{\lambda^2}{2} \phi_0 \left(\phi_0^2 - \nu^2 \right) = 0, \tag{31}$$

which has a domain wall solution, known as a "kink," in the following form [6]:

$$\phi_0(z) = \nu \tanh\left(\frac{\lambda \nu z}{2} \right). \tag{32}$$

Note that, when $\phi_0(z \to -\infty) = -\nu$ and $\phi_0(z \to +\infty) = \nu$. Therefore, the solution (32) separates the two lowest-energy states. This is why, this configuration is called a domain wall. **Figure 3** shows this behavior.

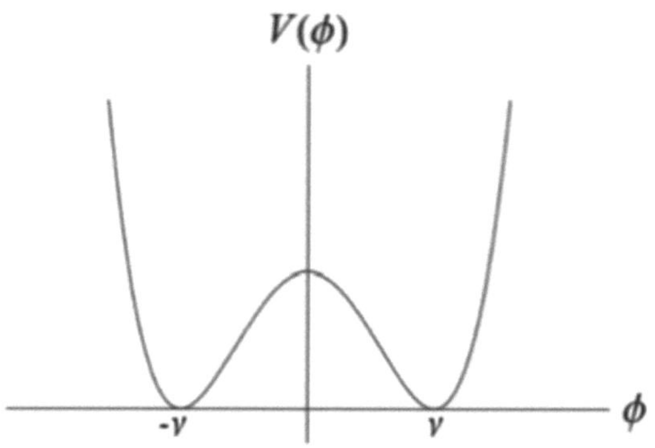

Figure 2.
Graph of the potential $V(\phi)$ in function of the field ϕ.

Fermions are particles with half-integer spin, such as the electron and proton, for example, the motion of these particles is described by the Dirac equation [8].

3.1.1 The location of the fermions

In Dirac's theory, fermions are described by spinors ψ, upon which the Dirac matrices act [8]. In five dimensions, the Dirac matrices Γ^{μ} can be defined based on the four-dimensional Dirac matrices γ^{μ} as follows:

$$\Gamma^{\mu} = \gamma^{\mu} \ , \Gamma^{z} = i\gamma^{5} \tag{33}$$

where $\gamma^5 = i\gamma^0\gamma^1\gamma^2\gamma^3$, and $\gamma^0, \gamma^1, \gamma^2$, and γ^3 are the Dirac matrices:

$$\gamma^5 = \begin{pmatrix} 0 & 1_{2\times2} \\ 1_{2\times2} & 0 \end{pmatrix} \tag{34}$$

We can verify that the Γ^5 matrices satisfy the usual Dirac matrix algebra:

$$\Gamma^A\Gamma^B + \Gamma^B\Gamma^A = 2g^{AB}. \tag{35}$$

The action of the Dirac field can be represented as follows:

$$S_{1/2} = \int d^4xdz\left(i\overline{\Psi}\Gamma^A\partial_A\psi - m\overline{\Psi}\Psi\right), \tag{36}$$

where the Lagrangian for the Dirac field in five dimensions is given by

$$\mathscr{L} = i\overline{\Psi}\Gamma^A\partial_A\Psi - m\overline{\Psi}\Psi. \tag{37}$$

Using the Euler-Lagrange equation, the Dirac equation in the context of extra dimensions takes the form:

$$\left(i\Gamma^A\partial_A - m\right)\Psi = 0 \tag{38}$$

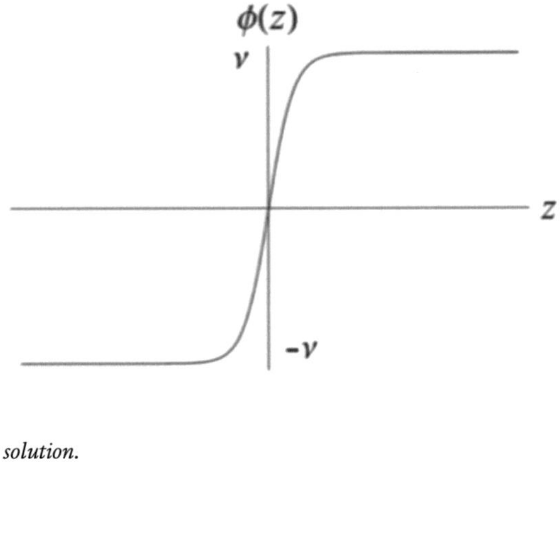

Figure 3.
Domain wall "kink" type solution.

This equation describes the free Dirac field. We will now consider the interaction of this field with a domain wall. As we saw in the previous section, the wall is described by a scalar field in a particular configuration. Let us consider a Yukawa-type interaction [7] between ψ and ϕ. The interaction action would be

$$S_{\mathrm{INT}} = -h\int d^4xdz\phi\overline{\Psi}\Psi, \tag{39}$$

where h is a coupling constant. Considering this, the total action of our system will consist of the sum of the purely fermionic action with the interaction action, $S = S_{1/2} + S_{\mathrm{INT}}$, with ϕ in the domain wall state ϕ_0. Thus, we have

$$S = \int d^4xdz\left(i\overline{\Psi}\Gamma^A\partial_A\psi - m\overline{\Psi}\Psi - h\phi_0\overline{\Psi}\Psi\right). \tag{40}$$

For a massless fermion, $m = 0$, the action reduces to

$$S = \int d^4xdz\left(i\overline{\Psi}\Gamma^A\partial_A\psi - h\phi_0\overline{\Psi}\Psi\right). \tag{41}$$

This gives us the following equation of motion:

$$i\Gamma^z\partial_z\Psi + i\Gamma^\mu\partial_\mu\Psi - h\phi_0\Psi = 0. \tag{42}$$

Now, our next step will be to solve the equation above. To do this, we will use the method of variable separation. We will consider $\Psi_{(5)} = \psi(x)f(z)$, where $\psi(x)$ is a 4-D Dirac spinor [9, 10] and $f(z)$ is a function depending only on the extra coordinate. Thus, we obtain

$$i\Gamma^z\psi(x)\frac{\partial_z f(z)}{f(z)} + i\Gamma^\mu\partial_\mu\psi(x) - h\phi_0\psi(x) = 0. \tag{43}$$

We impose that $\gamma^{(5)}\psi = \psi$. This does not imply any physical restriction since it asserts that massless particles have a well-defined chirality. Thus, we must have

$$\left[\frac{1}{f(z)}\partial_z f(z) + h\phi_0\right]\psi(x) - i\gamma^\mu\partial_\mu\psi(x) = 0. \tag{44}$$

Thus, to satisfy the condition above, we must have

$$i\gamma^\mu\partial^\mu\psi(x) = m\psi(x) \tag{45}$$

and simultaneously

$$\frac{df(z)}{dz} = (m - h\phi_0)f(z). \tag{46}$$

Due to (45), m is interpreted as the mass of the four-dimensional spinor. In the case where $m = 0$, we have the following configuration:

$$\frac{df(z)}{dz} = -h\phi_0 f(z). \tag{47}$$

Assuming $f(z)$ tends to zero at infinity, the solution of (47) is

$$f(z) = \exp\left(-h\int_0^z \phi_0 dz\right). \tag{48}$$

From the behavior of ϕ_0, see **Figure 3**, we observe that, for large $z, f(z)$ decreases exponentially. This means that the zero mode of the fermions is confined to the brane. Indeed, the wave function for zero-mode fermions is given by

$$\psi_0 = \exp\left(-\int_0^z h\phi_0(z)dz\right)\psi_L. \tag{49}$$

With this expression, we conclude that the zero mode is localized near $z = 0$, that is, on the domain wall. The wave function for large $|z|$ vanishes along the extra dimension.

With Eq. (46), it is possible to determine the admissible values for m. As already mentioned, m can be interpreted as the (KK) modes. According to Rubakov [6], the mass spectrum of the fermions confined to the brane is characterized by a gap between the zero mode and the other modes, which are proportional to $m_{(5)} = h\nu$. Thus, considering very large ν, the massive modes would be experimentally inaccessible.

3.2 Gravitational potential for an *n*-dimensional system

As mentioned earlier, in the ADD model, fields and matter are localized on the brane. However, the gravitational field can propagate along the extra dimension. To study the implications of this, we will derive the Newtonian gravitational potential in a space with D spatial dimensions generated by a body of mass m. For this purpose, two important considerations must be made. First, the gravitational field $\vec{g}$ must be conservative in a D -dimensional space. Second, the field equations should have the same form as the four-dimensional equation [11]. Thus, the equation will be

$$\vec{\nabla}\cdot\vec{g} = -4\pi G^D \rho, \tag{50}$$

where ρ is the matter density and G^D is the gravitational constant in D dimensions.

If the body has spherical symmetry, $\vec{g}(r)$ is radial and depends only on r. In this situation, it is useful to apply the divergence theorem to find the field produced by the body. To do this, consider a body of mass m and a sphere $S^{D-1}(r)$ of radius r, such that the sphere is centered on the mass m. This sphere, in turn, is the boundary of a ball $B^D(r)$ [11]. Integrating both sides of (50) over the ball $B^r(r)$, we have

$$\int_{B^{(D)}}\left(\vec{\nabla}\cdot\vec{g}\right)dv = -4\pi G^{(D)}\int_{B^{(D)}}\rho dv. \tag{51}$$

We know that $\int_{B^{(D)}}\left(\vec{\nabla}\cdot\vec{g}\right)dv \equiv \Phi_m$, meaning the integral equals the flux of $\vec{g}$ through $S^{D-1}(r)$. We need the volume of the sphere in a space with extra dimensions, which is given by

$$\mathrm{Vol}\left(S^{D-1}(r)\right) = r^{D-1}\frac{2\pi^{\delta/2}}{\Gamma(D/2)}. \tag{52}$$

Thus, considering that $\vec{g}(r)$ is conservative, $\vec{g}(r) = -\vec{\nabla}\phi$, where ϕ is the gravitational potential, and using the spherical symmetry of the body, we conclude that

$$\phi(r) = \frac{2\Gamma(D/2)}{\pi^{D/2-1}(2-2D)} \frac{Gm}{r^2}. \tag{53}$$

Note that if $D = 3$, the Newtonian potential is recovered. If $D = 4$, the potential ϕ is a function of $1/r^2$. This behavior is incompatible with observations of gravitational phenomena on long-distance scales, and therefore, it is necessary to compactify the extra dimension to make the model phenomenologically viable.

3.3 Length scale and the hierarchy problem

The constants c and $\hbar$ have been revealed by modern physics theories. In the study of gravitation, the gravitational constant G is also present. Using these three constants, we can construct the Planck unit system, whose basic units are Planck length l_P, Planck time t_P, and Planck mass m_P.We define these three units using these constants as follows:

$$l_P = \sqrt{\frac{G\hbar}{c^3}} = 1.61 \times 10^{-33}\text{cm} \quad , \quad t_P = \frac{l_P}{c} = \sqrt{\frac{G\hbar}{c^5}} = 5.4 \times 10^{-44}\text{s} \tag{54}$$

and

$$m_P = \sqrt{\frac{\hbar c}{G}} = 2.17 \times 10^{-5}\text{g}. \tag{55}$$

Considering that these quantities involve G associated with gravitation, $\hbar$ with quantum mechanics, and c with relativity, it is natural to think that these quantities define a scale at which quantum gravity effects may be significant.

We now aim to show a relationship between these quantities that applies to a D-dimensional space. In D dimensions, the gravitational constant G^D has the dimension of G, and this is expressed as

$$(l_D)^{D-2} = \frac{G^{(D)}\hbar}{c^3}. \tag{56}$$

From this equation, we can see that the value of l_P would be an effective value derived from the fundamental constant l_D, defined in the ambient space [3, 11]. In a space containing an extra compact dimension with the topology of a circle of radius l, whose dimension length is $2\pi l$, the gravitational constants are related as follows:

$$\frac{G^{(5)}}{G^{(4)}} = 2\pi l = l_c, \tag{57}$$

where l_c is the length of the extra dimension. The gravitational constants of ordinary 4-D spacetime and five-dimensional space (one compact dimension) differ by a factor proportional to the length of the extra dimension. We can generalize the above expression for an n-dimensional space using the relation:

$$\frac{G^{(D)}}{G^{(4)}} = (2\pi l)^D = (l_c)^{D-2}. \tag{58}$$

This assumes that the length of the extra dimension is the same. Combining (56) with the above equation, we can express l_c in terms of l_P in a 4-D space and the Planck length in D-dimensions (l_D):

$$l_c = l_D \left(\frac{l_D}{l_c}\right)^{\frac{2}{D-4}}. \tag{59}$$

The hierarchy problem would be resolved if we admitted that the Planck length l_P in a space with D extra dimensions was of the same order as the electroweak scale $l_D \approx 10^{-18}$ cm. With this condition, we obtain a relationship for the size of the extra dimension l_c and the number of dimensions of space:

$$l_c = 10^{-18}\left(10^{15}\right)^{\frac{2}{D-4}}. \tag{60}$$

From the above relation, we observe that, in the case D = 5, that is, one extra dimension, $l_c \sim 10^{12}$ cm. Thus, a universe with only one extra dimension would not be compatible with this model, as such a dimension would already have been observed in nature. On the other hand, considering a universe with six dimensions (two extra dimensions), we would have $l_c \sim 0.01$ mm. This is a distance at which Newtonian gravitational laws are currently being tested.

We conclude, therefore, that possible deviations from the inverse-square gravitational law could be interpreted as evidence for extra dimensions. Note that the greater the number of extra dimensions, the smaller the length of the extra dimension. For the case of $D = 10$, that is, six extra dimensions, as suggested by the string theory [12], the size would be $l_c \approx 10^{-13}$ cm, which is still much larger than the Planck length. Hence, these models are also known as long-scale extra dimension models.

4. The Randall-Sundrum models

The Randall-Sundrum (RS) models [4, 5] are also brane-world models. Compared to the ADD model, the (RS) models stand out for showing that the five-dimensional ambient space is not incompatible with experience. This is possible due to the ambient space has a negative cosmological constant whose value is adjusted to the brane tension.

There are two models, known as RSI and RSII. In the RSI model, there are two branes separated in the extra dimension, which is compact and finite. In the RSII model, there is only a single brane, allowing the extra dimension to have infinite length.

4.1 The RSI model

In the ambient space of the RSI model [4], there are two branes, making the extra dimension compact. The branes are located at two fixed points in the ambient space. The first brane has positive tension σ_+ at $z = 0$, and the second brane has negative tension σ_-at$z = z_c$, where z_c is the length of the extra dimension.

The extra dimension is assumed to have the topology of the space S^1/Z_2, described as follows. Let $\theta \in [-\pi, \pi]$ be the coordinate labeling the points of the circle S^1. The topology S^1/Z_2 is obtained by identifying the points $-\theta$ and θ. Note that there are two fixed points with respect to this identification: $\theta = 0$ and $\theta = \pi$. The branes are located at these points.

The background metric for the RSI model is given by [13]

$$ds^2 = a^2(z)\eta_{\mu\nu}dx^\mu dx^\nu - dz^2, \tag{61}$$

where $\eta_{\mu\nu}$ is the Minkowski metric in (1 + 3)-dimensional spacetime, and $a(z)$ is the so-called warp factor, given by

$$a(z) = e^{-\kappa|z|}, \tag{62}$$

where the constant κ is determined by solving Einstein's equations. The behavior of $a(z)$ in S^1/Z_2 is illustrated below.

Now, let us consider the field equations for the ambient space to determine the value of κ. Einstein's equations in the ambient space are

$$R_{AB} - \frac{1}{2}g_{AB}R = T_{AB}, \tag{63}$$

where T_{AB} is the energy-momentum tensor of the source. It includes a term for the cosmological constant Λ and the brane tension [4, 12].

$$T_{AB} = \Lambda g_{AB} + 8\pi G_{(5)}\tau_{AB}, \tag{64}$$

where τ_{AB} represents the brane's energy content and is given by

$$\tau_{AB} = \begin{cases} T_{\mu\nu} = \sigma g^{(5)}_{\mu\nu}\delta(z) - \sigma g^{(5)}_{\mu\nu}\delta(z - z_c), \\ \qquad\qquad\qquad\qquad T_{Az} = 0. \end{cases} \tag{65}$$

The expression (65) implies that there is no energy flux between the ambient space and the brane. Using the metric (61) and the energy-momentum tensor (64), the Einstein equations are

$$\left[-3\left(\frac{a'}{a}\right)^2 - 3\left(\frac{a''}{a}\right)\right] = 8\pi G_{(5)}[\sigma\delta(z) - \sigma\delta(z - z_c)] + \Lambda, \tag{66}$$

$$G_{\mu z} = 0, \tag{67}$$

$$\left[-6\left(\frac{a'}{a}\right)\right] = \Lambda. \tag{68}$$

The Eq. (67) is automatically satisfied, while Eq. (68) holds if

$$\kappa^2 = -\frac{1}{6}\Lambda, \tag{69}$$

which requires the cosmological constant to be negative for the metric (61) to solve Einstein's equations. Examining Eq. (66) and integrating near $z = 0$ over the interval $-\epsilon < z < \epsilon$, and taking the limit $\epsilon \to 0$, we find that the brane tension is given by

$$\sigma^2 = -\frac{3}{32\pi^2 G_{(5)}^2}\Lambda. \tag{70}$$

Note that we assume the brane has positive tension. If we take the solution at $z = z_c$, the result will be the same.

4.1.1 The mass spectrum for gravitons

The process of linearizing gravity in the context of brane-world scenarios leads to determining the mass spectrum for gravitons. In this framework, a perturbation $h_{\mu\nu}$ in the metric is considered, whose temporal evolution results in the following differential equation:

$$h''_{\sigma\rho} - 4\kappa^2 h_{\sigma\nu} - a^{-2}\partial^\mu\partial_\mu h_{\sigma\nu} = 0, \tag{71}$$

with κ defined appropriately (69). To solve the equation above, we will again use the method of variable separation. For this, we consider a solution of the type $h_{\sigma\nu} = \Psi(z)\Phi_{\sigma\nu}(x)$. Substituting this expression into (71), we obtain

$$\Phi_{\sigma\nu}(x)\partial_z^2\Psi(z) - \Psi(z)a^{-2}\Box\Phi_{\sigma\nu}(x) - 4\kappa^2[\Psi(z)\Phi_{\sigma\nu}(x)] = 0, \tag{72}$$

which leads to

$$\Psi''(z) - 4\kappa^2\Psi(z) + C\Psi(z) = 0, \tag{73}$$

$$\Box\Phi_{\sigma\nu} = -C\Phi_{\sigma\nu}. \tag{74}$$

We will see that it is possible to show that the constant C can take positive values. Thus, $C = m^2$, which allows us to rewrite (74) as

$$\Box\Phi_{\sigma\nu}(x) + m^2\Phi_{\sigma\nu}(x) = 0. \tag{75}$$

We observe that this equation has the form of the Klein-Gordon equation for the field $\Phi_{\sigma\nu}(x)$ with mass m. Since the field $\Phi_{\sigma\nu}(x)$ is a function of the brane coordinates, we can interpret it as a graviton with mass m, that is, a linear perturbation of the metric, for observers confined to the brane. Thus, for each allowed m, we have a field $\Phi_{\sigma\nu}(x)$, referred to as a Kaluza-Klein mode. Eq. (73), in turn, takes the form:

$$\Psi''(z) - 4\kappa^2\Psi(z) + \frac{m^2}{a^2}\Psi(z) = 0. \tag{76}$$

For $m = 0$, we have a particular solution of the form:

$$\Psi(z) = C_0 e^{-2\kappa z}. \tag{77}$$

This solution is known as the zero mode $m = 0$. Later, we will see that it is fundamentally important for recovering the four-dimensional behavior of the gravitational field. If m $\neq 0$, we can make a coordinate transformation $x = \frac{m}{\kappa}e^{\kappa z}$. In this case, Eq. (76) becomes

$$\frac{d^2\Psi(x)}{dx^2} + \frac{1}{x}\frac{d\Psi(x)}{dx} + \left(1 - \frac{4}{x^2}\right)\Psi(x) = 0. \tag{78}$$

This equation is a Bessel equation for $n = 2$, whose general solution has the form:

$$\Psi(z) = AJ_2\left(\frac{m}{\kappa}e^{\kappa z}\right) + BN_2\left(\frac{m}{\kappa}e^{\kappa z}\right), \tag{79}$$

where $J_2(z)$ and $N_2(z)$ are the Bessel functions of the first and second kinds, respectively, of order 2, the following equation:

$$\Psi(x) = C_m\left[J_1\left(\frac{m}{\kappa}\right)N_2\left(\frac{m}{\kappa}e^{\kappa z}\right) - N_1\left(\frac{m}{\kappa}\right)J_2\left(\frac{m}{\kappa}e^{\kappa z}\right)\right], \tag{80}$$

where C_m is a normalization constant. For $z = z_c$, we have

$$\frac{J_1\left(\frac{m}{\kappa}e^{\kappa z_c}\right)}{N_1\left(\frac{m}{\kappa}e^{\kappa z_c}\right)} = \frac{J_1\left(\frac{m}{\kappa}\right)}{N_1\left(\frac{m}{\kappa}\right)}. \tag{81}$$

The possible values of m are the roots of this equation, that is, the spectrum of gravitons is provided by the set of solutions to this equation [14]. For the first solutions of (81), we have

$$m \simeq \kappa e^{-\kappa z_c}. \tag{82}$$

Note that the mass of the gravitons depends on the size of the extra dimension. In this case, if we choose a large κ and a small z_c such that $z_c\kappa > 1$, we observe that the mass of the first (KK) mode is very large $(m \sim \kappa)$. Thus, exciting the first massive mode would require an enormous amount of energy, which is beyond the reach of experiments. This explains why massive gravitons have not been detected. Therefore, a spacetime with just one extra dimension would be compatible with observational data at the energy scale of current particle accelerators.

4.2 The RSII model

We know that, in the RSI model, there are two branes: one with positive tension and another with negative tension. Now, for the RSII model [5], suppose we remove the brane with negative tension and send it to an infinite distance. This causes the extra dimension to no longer be compact and instead have infinite length. One direct consequence of this is the absence of a discrete mass spectrum.

Under these circumstances, it would be expected that gravitons with small mass would be accessible at low energy levels, which would bring the theory into conflict with experience. However, we will see later that the existence of a normalizable zero mode allows us to recover four-dimensional behavior over long distances.

4.2.1 The normalizable zero mode and continuous Kaluza-Klein modes

Let us return to Eq. (73), as we have seen that the KK modes satisfy this equation. To study the normalization of $\Psi(z)$, we will use Sturm-Liouville theory. In this theory, eigenfunctions associated with discrete eigenvalues are normalizable, while eigenfunctions corresponding to continuous eigenvalues are normalizable in the distributional sense, that is, according to the Dirac delta.

Assuming the eigenfunctions are finite for $z \to \infty$, it is possible to show that all eigenvalues are non-negative [15]. Another important fact is the existence of a normalizable zero mode. Indeed, if we consider $m = 0$ in Eq. (76), we have

$$\Psi_0(x) = C_0 e^{-2\kappa|z|} \tag{83}$$

It is a fundamental point of this model that ψ_0 is normalizable even for a non-compact dimension, i.e., $-\infty < z < \infty$. In fact, we can impose

$$\int_{-\infty}^{+\infty} |\Psi_0(z)|^2 dz = 1 \tag{84}$$

which gives us $C_0 = \sqrt{2\kappa}$. For the KK modes, the normalized functions are

$$\Psi_m(z) = \sqrt{\frac{m}{\kappa}} \frac{\left[J_1\left(\frac{m}{\kappa}\right)N_2\left(\frac{m}{\kappa}e^{\kappa z}\right) - N_1\left(\frac{m}{\kappa}\right)J_2\left(\frac{m}{\kappa}e^{\kappa z}\right)\right]}{\sqrt{J_1\left(\frac{\kappa}{m}\right)^2 + N_1\left(\frac{\kappa}{m}\right)^2}}. \tag{85}$$

4.2.2 The Green's function

To study the behavior of the gravitational field, the use of the Green's function is necessary. If a specific event occurs at a point $\vec{x}'$ in space at a time t' (in our case on the brane), the Green's function provides the influence of that event at a point $\vec{x}$ at time t. Considering the behavior of the linearized gravitational field, we must study the inhomogeneous solution:

$$\left[\partial_z^2 - 2\left(2\kappa^2 - 2\kappa\delta(z)\right) + a^{-2}\Box^{(4)}\right] h_{\mu\nu} = \Sigma_{\mu\nu}. \tag{86}$$

The new term $\Sigma_{\mu\nu}$ acts as a source for $h_{\mu\nu}$. Thus, according to the Green's formalism, the solution of (86) can be expressed as

$$h_{\mu\nu} = \int dz' d^4x' G(x, z; x', z') \tag{87}$$

where $G(x, z; x', z')$ is the Green's function, satisfying the equation:

$$\left[\partial_z^2 - 2\left(2\kappa^2 - 2\kappa\delta(z)\right) + a^{-2}\Box^{(4)}\right] G(x, z; x', z') = \delta^4(x - x')\delta(z - z'). \tag{88}$$

From this equation, we can construct the Green's function using the method of separation of variables, which leads us to

$$\Box\Phi_{\sigma\nu} - m^2\Phi_{\sigma\nu} = 0 \tag{89}$$

$$\Psi''(z) - 2\left(2\kappa^2 - 2\kappa\delta(z)\right)\Psi(z) + \frac{m^2}{a^2}\Psi(z) = 0 \tag{90}$$

The Sturm-Liouville theory provides us with a very important result. For functions, the complete set of eigenfunctions forms a complete basis that satisfies the same boundary conditions as the eigenfunctions [16]. For Eq. (89), the eigenfunctions are plane waves of the form $\frac{1}{(2\pi)^4}e^{-k_\mu x^\mu}$, with $k_\mu k^\mu = m^2$. For Eq. (90), the eigenfunctions are given by

$$u_0(z) = \sqrt{2\kappa}e^{-\kappa|z|} \tag{91}$$

$$u_m(z) = \sqrt{\frac{m}{\kappa}} \frac{\left[J_1\left(\frac{m}{\kappa}\right)N_2\left(\frac{m}{\kappa}e^{\kappa z}\right) - N_1\left(\frac{m}{\kappa}\right)J_2\left(\frac{m}{\kappa}e^{\kappa z}\right)\right]}{\sqrt{J_1\left(\frac{\kappa}{m}\right)^2 + N_1\left(\frac{\kappa}{m}\right)^2}} \tag{92}$$

It is then seen that the functions $\frac{1}{(2\pi)^4}e^{-k_\mu x^\mu}u_m(z)$ constitute a basis for functions defined in five dimensions. In this basis, the Green's function can be expressed as

$$G(x,z;x',z') = \int \frac{d^4k}{(2\pi)^4} b_{0k}u_0(z)e^{ikx} + \int \frac{d^4k}{(2\pi)^4} dm, b_{mk}u_m(z)e^{ik_\mu x^\mu}. \tag{93}$$

The coefficients b_{mk} and u_m can be determined via Eq. (88), from which it is possible to show that the time-independent Green's function [5] is
given by

$$G(x,z;x',z') = \left(-\frac{1}{4\pi}\frac{1}{R}\right)\kappa e^{-2\kappa(z+z')} + \int_0^\infty dm, u_m(z')\left(-\frac{1}{4\pi}\frac{e^{-mR}}{R}\right), \tag{94}$$

where $R = \ |\vec{x} - \vec{x}'|$. Note that, because the zero mode $m = 0$ is normalizable, we recover the Newtonian case where the potential decreases inversely with distance. For the massive modes, the Green's function decreases exponentially with distance.

Now, suppose we have a particle localized on the brane at $z = 0$ [4]. The Green's function evaluated at another point on the brane ($z' = 0$) is

$$G(x,x') = -\frac{\kappa}{4\pi R} - \frac{1}{4\pi R}\int_0^\infty dm, [u_m(0)]^2 e^{-mR}, \tag{95}$$

where $u_m(0)$ are the eigenstates evaluated at $z' = 0$. Thus, if we want to calculate the gravitational potential for a point far from the brane, we assume $R > \kappa^{-1}$. Since in (95) the exponential is decreasing, the integral is significant only for $m < m^*$, where m^* is proportional to $\frac{1}{R}$.

$$mR < m^*R \sim 1 \tag{96}$$

From these conditions, it follows that, in the integral in (95), we can consider $m \ll \kappa$. This allows us to expand the Bessel functions for small arguments. It can be shown that, to first approximation, for m/κ, we have

$$[u_m(0)]^2 = \frac{m}{\pi^2\kappa}. \tag{97}$$

Substituting (97) into (95), we find the Green's function in the following form:

$$G(x,z;x',z') = -\frac{\kappa}{4\pi R}\left[1 - \frac{1}{\pi^2\kappa^2R^2}\right]. \tag{98}$$

This Green's function represents the gravitational potential of a massive object located on the brane. We observe that the Green's function due to the zero mode provides the four-dimensional gravitational potential at long distances [5]. The KK modes provide a correction to the potential.

5. Conclusions

In this chapter, it was possible to analyze the main theories of extra dimensions as well as their fundamental importance for the construction of unification theories, such as string theory, for example. We also observed the experimental difficulties in detecting these dimensions, considering their extremely small scale. It is expected that, over the years, with technological advancements, particularly in atomic spectroscopy and specifically in neutron interferometry techniques, evidence of extra dimensions in our Universe may be detected.

Conflict of interest

The author declare no conflict of interest.

Abbreviations

GRT	General Relativity Theory
GUT	Grand Unifity Theory
KK	Kaluza-Klein

Author details

Eugênio Maciel
Universidade Federal de Campina Grande, Sumé, Brazil

*Address all correspondence to: eugenio.bastos@professor.ufcg.edu.br

References

[1] Theodor K. Zum Unitätsproblem in der Physik. Sitzungsber. Sitzungsberichte der Preussischen Akademie der Wissenschaften zu Berlin. (Math. Phys.). 1921;**1921**:966-972

[2] Klein O. Zeitschrift für Physik. Elektrodynamik und wellenmechanik vom standpunkt des korrespondenzprinzips. 1927;**41**: 407-422. DOI: 10.1007/BF01400205

[3] Arkani-Hamed N, Dimopoulos S, Dvali G. The hierarchy problem and new dimensions at a millimeter. Physics Letters B. 1998;**429**(3–4):263-272. DOI: 10.1016/S0370-2693(98)00466-3

[4] Randall L, Sundrum R. Large mass hierarchy from a small extra dimension. Physical Review Letters. 1999;**83**(17): 3370. DOI: 10.1103/PhysRevLett.83.3370

[5] Randall L, Sundrum R. An alternative to compactification. Physical Review Letters. 1999;**83**(23):4690. DOI: 10.1103/ PhysRevLett.83.4690

[6] Rubakov VA, Shaposhnikov ME. Extra space-time dimensions: Towards a solution to the cosmological constant problem. Physics Letters B. 1983;**125**(2–3):139-143. DOI: 10.1016/0370-2693(83) 91254-6

[7] Rubakov VA. Large and infinite extra dimensions. Physics-Uspekhi. 2001; **44**(9):871. DOI: 10.1070/ PU2001v044n09ABEH001000

[8] Dahia F, Romero C. Confinement and stability of the motion of test particles in thick branes. Physics Letters B. 2007;**651** (2–3):232-238. DOI: 10.1016/j. physletb.2007.06.004

[9] Greiner W. Relativistic Quantum Mechanics. Vol. 2. Berlin: Springer; 2000. ISSN:978-3-540-99535-7

[10] Bjorken JD, Drell SD, Mansfield JE. Relativistic Quantum Mechanics. New York: McGraw-Hill College; 1965. pp. 81-82. DOI: 10.1063/1.3047288

[11] Zwiebach B. A First Course in String Theory. Cambridge: Cambridge University Press; 2004. DOI: 10.1016/ 0550-3213(86)90177-X 26

[12] Shifman M. Large extra dimensions: Becoming acquainted with an alternative paradigm. International Journal of Modern Physics A. 2010;**25**(02n03): 199-225. DOI: 10.1142/ S0217751X10048548

[13] Brown LS, Kaku M, Greenberg OW. Quantum Field Theory and Quantum Field Theory: A Modern Introduction1994. pp. 104-106. DOI: 10.1063/1.2808409

[14] Arfken G, Romain JE. Mathematical methods for physicists. Physics Today. 1967;**20**(5):79-79. DOI: 10.1063/ 1.3034326

[15] Giddings SB, Katz E, Randall L. Linearized gravity in brane backgrounds. Journal of High Energy Physics. 2000; **2000**(03):023. DOI: 0.1088/1126-6708/ 2000/03/023

[16] Morse PM, Feshbach H, Condon EU. Methods of Theoretical Physics Parts I and II. Vol 1. New York: McGraw-Hill College; 1954. pp. 15-16. DOI: 10.1063/ 1.3061460

Chapter 4

Thermodynamics of a Supersymmetric Oscillator

Jishad Kumar

Abstract

The thermodynamic properties of a supersymmetric (SUSY) oscillator are analyzed within the framework of the canonical ensemble. Analytical expressions for the heat capacity and entropy are derived and subsequently studied in their asymptotic limits. It is demonstrated that the specific heat of the SUSY oscillator adheres to the third law of thermodynamics at low temperatures and satisfies the classical equipartition theorem in the high-temperature regime. This occurs despite the presence of a fermionic component, represented as a two-level system, within the SUSY oscillator. Notably, the specific heat of the fermionic subsystem vanishes at both zero and very high temperatures. The heat capacity of the SUSY oscillator exhibits a distinct peak at an intermediate temperature as the system approaches the low-temperature limit, a behavior reminiscent of that observed in a fermionic (two-level) system. Furthermore, the case of an electron in a magnetic field is examined in detail, revealing the supersymmetric nature of the system when the spin of the electron is incorporated into the analysis. The spontaneous breaking of supersymmetry in the Landau model is discussed in the context of an additional confinement potential, characterized by a harmonic oscillator frequency.

Keywords: supersymmetry, quantum mechanics, oscillator, Landau model, Fock-Darwin model, thermodynamics

1. Introduction

Thermodynamics encapsulates the fundamental properties that physical systems exhibit in thermal equilibrium. Established in the nineteenth century by pioneers such as Robert Mayer and Hermann Helmholtz, it introduced the principle of energy conservation, which encompasses the equivalence of work and heat [1]. The framework of thermodynamics is founded on its core concepts, entropy, temperature, and the three laws governing state variables, which allow an independent and comprehensive description of system dynamics. Unlike classical mechanics and electrodynamics, which have evolved substantially over the past century, thermodynamics has maintained its foundational structure and robustness. The interplay of thermodynamics with statistical physics introduces an intriguing dimension, particularly for microscopic systems. Statistical physics bridges the macroscopic and microscopic realms, deriving emergent macroscopic properties from atomic structures and

IntechOpen

microscopic dynamics, such as temperature, pressure, and material constants. These emergent properties are hallmarks of many-body systems, arising from the interactions of numerous particles (atoms or molecules), and are typically absent in individual microscopic systems. Statistical physics offers powerful methods to predict and analyze properties like specific heat by linking microscopic parameters, such as interparticle interaction energies, to macroscopic thermodynamic quantities.

Incorporating probabilistic frameworks into physics, pioneered by statistical mechanics, finds its foundation in the fundamental concept of entropy. The term *entropy* was formally introduced by Clausius in 1865 [2] and later extended through Maxwell's innovative "*Gedankenexperiment*" in 1878 [3]. Maxwell also proposed the term *Statistical Mechanics*. The unification of thermodynamics and statistical physics culminated in the development of statistical thermodynamics, epitomized by the Boltzmann-Planck formula for entropy [4], $S = k_B \ln W$, W representing the thermodynamic probability), and Boltzmann's ergodic hypothesis [5], independently refined by Josiah Willard Gibbs [6], the progenitor of ensemble theory. In classical mechanics, the evolution of a system in phase space is deterministically defined once the initial conditions of its coordinates and momenta are specified. However, thermodynamics and statistical physics focus not on tracking specific trajectories but on understanding averaged behaviors over extended timescales and spatial regions. These timescales, much longer than the microscopic state transitions, render direct trajectory-based calculations infeasible because of the sheer complexity arising from interparticle interactions and environmental influences. Statistical mechanics solves this challenge through ensembles: collections of identically prepared systems, each occupying an allowed state of the evolving system. The probability distribution of states within an ensemble mirrors the probabilities observed along the actual trajectory of the system. This equivalence enables ensemble averaging to replace time averaging, thereby simplifying calculations. The fundamental assumption that ensemble and time averages yield identical results underpins the principle of ergodicity, a cornerstone of statistical mechanics.

The advent of quantum theory, introduced by Max Planck, marked a paradigm shift in physics, fundamentally transforming its underlying principles. This revolutionary development inevitably influenced classical statistical mechanics, which underwent significant restructuring to incorporate quantum concepts. A key innovation was the principle of indistinguishability of identical particles, which necessitated a complete overhaul of the formalism of statistical mechanics. Despite these profound changes, the foundational concept of ergodicity remained intact in the quantum domain. The microcanonical and canonical ensembles persisted with minor adjustments, while the grand canonical ensemble emerged as a novel addition to the formalism. The probabilistic nature of wave mechanics added a new dimension to the statistical framework. In classical statistical mechanics, the statistical character arises primarily from the system's many particles. In quantum mechanics, this is augmented by the probabilistic interpretation of quantum states. Together, these aspects led to a comprehensive reformulation of ensemble theory, culminating in introducing the density matrix—a quantum mechanical counterpart of the classical phase-space density function. The concept of the density matrix was first introduced by Landau [7], subsequently formalized by von Neumann [8], and later refined by Dirac [9–11]. The density matrix not only preserves the statistical essence of classical mechanics but also, encapsulates the inherently probabilistic nature of quantum systems, establishing a robust framework for studying quantum statistical mechanics. This reformulation allowed for the seamless integration of quantum phenomena into the statistical description of physical systems, bridging the gap between classical and quantum regimes.

Building on these preliminary historical insights, this chapter analyzes the statistical thermodynamics of a supersymmetric (SUSY) oscillator, a simplified yet insightful model derived from quantum field theory. Before delving into the SUSY oscillator, we begin with a detailed review of the well-established bosonic and Fermionic oscillators. The thermodynamic properties of these systems, particularly their specific heat and entropy, are examined comprehensively, with careful attention given to their behavior in both the low- and high-temperature limits. The chapter starts with an introduction to the thermodynamic quantities of interest, providing the foundation for the following discussions. Next, the bosonic and Fermionic oscillators are analyzed in detail, and their thermodynamic properties are thoroughly explored. Building on this groundwork, we transition to studying the supersymmetric oscillator, examining its thermodynamic behavior and highlighting the distinctions introduced by supersymmetry. Subsequently, we discuss a real-world example of supersymmetry in the context of an electron in a magnetic field, incorporating the effects of Zeeman splitting. This example illustrates the manifestation of supersymmetry in a physical system and its impact on observable quantities. The chapter addresses two cases where supersymmetry is broken, offering insights into the mechanisms and implications of SUSY breaking in physical systems. The chapter concludes with a summary of the key findings and their broader impact, and the references follow thereafter. Through this systematic exploration, the chapter aims to provide a comprehensive understanding of the thermodynamic behavior of the supersymmetric oscillator and its relevance to real-world systems, emphasizing the profound influence of supersymmetry in statistical thermodynamics.

2. Thermodynamic quantities of interest

Within the framework of the canonical ensemble, most practical scenarios involve controlling the temperature of a system rather than its energy. The canonical ensemble describes a system with a fixed number of particles that can exchange energy with its surrounding environment, often referred to as a heat bath. When a system is in thermal equilibrium with a much larger heat bath, it is maintained at a fixed temperature. In such cases, the system's energy is not constant but instead fluctuates around a well-defined average value. The magnitude of these fluctuations is small compared to the average energy of the system. Consequently, probabilities for states with different energies need to be determined. The Boltzmann factor governs these probabilities, e^{-E/k_BT}, where E is the energy of a given microscopic state, k_B is the Boltzmann constant, and T is the absolute temperature.

The system is assigned a temperature in the canonical ensemble, and all accessible states are allowed to occur with probabilities weighted by the Boltzmann factor. This leads to introducing the partition function, an ensemble-averaged quantity central to connecting statistical mechanics to thermodynamics. The partition function is defined as:

$$\mathcal{Z} = \sum_{\text{all microscopic states}} e^{-\beta E(\text{microscopic states})}, \tag{1}$$

where $\beta = 1/(k_BT)$. While the partition function does not have a direct physical interpretation, it is the foundation for deriving thermodynamic properties.

Once the partition function is known, various thermodynamic quantities can be computed. The free energy, which is a fundamental thermodynamic potential, is given by:

$$F = -k_B T \ln \mathcal{Z}. \tag{2}$$

The internal energy, or the ensemble-averaged energy, is obtained as:

$$\mathcal{U} = -\frac{\partial}{\partial \beta} \ln \mathcal{Z}. \tag{3}$$

The entropy of the system is expressed as:

$$S = k_B \left(\ln \mathcal{Z} - \beta \frac{\partial}{\partial \beta} \ln \mathcal{Z} \right). \tag{4}$$

Finally, the specific heat at constant volume is derived from the internal energy or entropy as:

$$C_V = -k_B \beta^2 \frac{\partial \mathcal{U}}{\partial \beta}, \tag{5}$$

or equivalently:

$$C_V = -\beta \frac{\partial S}{\partial \beta}. \tag{6}$$

These equations provide the necessary tools for calculating thermodynamic properties, including specific heat, which is a key observable in understanding energy distribution among the various modes of a system. Specific heat measurements are particularly valuable, as their variation with temperature reveals insights into how energy is partitioned among a solid's vibrational, rotational, and other energy-absorbing modes. Additionally, these measurements help correlate other material properties, such as thermal expansion, with the underlying lattice structure and vibrations. Specific heat at constant volume, C_V, is especially significant because of its direct connection to the internal energy, offering a microscopic perspective on the system's structure.

Thermodynamics provides a macroscopic description of actual physical systems in thermal equilibrium with a heat bath. Its simplicity and robustness lie in its ability to describe the dynamics of physical systems independently through quantities such as entropy, temperature, and the three fundamental laws of thermodynamics [12]. However, thermodynamics attains greater depth and versatility when combined with statistical mechanics. Statistical mechanics bridges the microscopic and macroscopic worlds by deriving emergent macroscopic properties from a system's atomic structure and microscopic dynamics. These emergent properties, such as temperature and pressure, are characteristic of many-body systems and typically do not manifest in isolated microscopic systems. By connecting microscopic information, such as particle interactions, to macroscopic properties, statistical mechanics enhances thermodynamics' explanatory and predictive power, leading to the discipline of statistical thermodynamics.

This paper focuses on the thermodynamics of a supersymmetric (SUSY) oscillator. The simple harmonic oscillator is one of the most fundamental and extensively studied models in physics, owing to its exact solvability and elegant mathematical structure. This model underpins the explanation of numerous phenomena across various domains of physics and other scientific disciplines. The supersymmetric harmonic oscillator extends the standard model by incorporating both bosonic and fermionic variables, serving as a prototypical example of supersymmetry and its associated algebra [13, 14]. Supersymmetry, a theoretical framework interchanging bosons and fermions, is a novel spacetime symmetry significantly influences quantum field theory. In supersymmetric quantum mechanics, the Hamiltonian of a system remains invariant under the transformation of bosonic into fermionic variables and vice versa. The relevance of supersymmetry extends beyond quantum mechanics to high-energy physics and cosmology. It plays a vital role in string theory, where it was first introduced, and has since been integrated into supergravity theories [15]. Supersymmetry, commonly abbreviated as SUSY, offers profound insights into unresolved questions in particle physics, string theory, and cosmology, establishing itself as a cornerstone of modern theoretical physics. This chapter explores the statistical thermodynamics of the SUSY oscillator, aiming to elucidate its rich mathematical structure and physical implications within the broader context of supersymmetric quantum mechanics.

3. Bosonic and fermionic oscilators

The Hamiltonian of a one-dimensional quantum harmonic oscillator with a natural vibrational frequency ω can be elegantly expressed using the creation ($a_B^\dagger$) and annihilation (a_B) operators, which satisfy specific commutation relations. The subscript B is introduced to denote "Bosonic" operators for clarity, as analogous operators will be defined later for the fermionic case. In terms of these operators, the Hamiltonian of the Bosonic oscillator is written as [16]:

$$\mathcal{H}_B = \hbar\omega\left(a_B^\dagger a_B + \frac{1}{2}\right) = \hbar\omega\left(N_B + \frac{1}{2}\right), \tag{7}$$

where $N_B = a_B^\dagger a_B$ is the number operator. The creation and annihilation operators obey the commutation relation

$$\left[a_B, a_B^\dagger\right]_- = 1, \tag{8}$$

and act on the eigenstates $|n_B\rangle$ of the number operator as follows:

$$a_B^\dagger|n_B\rangle = \sqrt{n_B+1}|n_B+1\rangle, \tag{9}$$

$$a_B|n_B\rangle = \sqrt{n_B}|n_B-1\rangle. \tag{10}$$

The eigenstates of the number operator are given by:

$$|n_B\rangle = \frac{1}{\sqrt{n_B!}}\left(a_B^\dagger\right)^{n_B}|0\rangle_B, \tag{11}$$

where $|0\rangle_B$ is the vacuum state, satisfying $a_B|0\rangle_B = 0$. When the Hamiltonian acts on the eigenstates, the energy eigenvalues are given by:

$$\mathcal{H}_B|n_B\rangle = \left(n_B + \frac{1}{2}\right)\hbar\omega|n_B\rangle, \tag{12}$$

with corresponding energy eigenvalues:

$$E_n = \left(n_B + \frac{1}{2}\right)\hbar\omega. \tag{13}$$

These eigenvalues are quantized and separated by intervals of $\hbar\omega$, with a nonzero minimum energy $E = \frac{1}{2}\hbar\omega$, referred to as the *zero-point energy*. The symmetric nature of the Hamiltonian reflects the bosonic statistics, where the associated operators satisfy commutation relations.

For fermionic systems, which inherently exhibit antisymmetry, the operators obey anti-commutation relations instead of commutation relations. The Hamiltonian of a fermionic oscillator with frequency ω is defined as Ref. [17]:

$$\mathcal{H}_F = \frac{\hbar\omega}{2}\left(a_F^\dagger a_F - a_F a_F^\dagger\right), \tag{14}$$

where the subscript F denotes "Fermionic" operators. If the operators a_F and $a_F^\dagger$ were to obey commutation relations similar to the bosonic case, the Hamiltonian would reduce to:

$$\mathcal{H}_F = -\frac{\hbar\omega}{2}, \tag{15}$$

implying no dynamics. To account for the inherent antisymmetry of fermions, the operators are instead defined to satisfy anti-commutation relations:

$$[a_F, a_F]_+ = 0, \tag{16}$$

$$\left[a_F^\dagger, a_F^\dagger\right]_+ = 0, \tag{17}$$

$$\left[a_F, a_F^\dagger\right]_+ = 1. \tag{18}$$

These anti-commutation relations imply that the system follows Fermi-Dirac statistics, which is reflected in the eigenvalues of the number operator:

$$N_F = a_F^\dagger a_F, \quad N_F^2 = N_F, \quad N_F(N_F - 1) = 0. \tag{19}$$

This indicates that the eigenvalues of N_F are restricted to 0 or 1, consistent with the Pauli exclusion principle. The Hamiltonian of the fermionic oscillator then takes the form:

$$\mathcal{H}_F = \hbar\omega\left(N_F - \frac{1}{2}\right). \tag{20}$$

If $|n_F\rangle$ is an eigenstate of N_F, then:

$$N_F|n_F\rangle = n_F|n_F\rangle, \quad n_F = 0, 1. \tag{21}$$

The ground state $|0\rangle$, with no quanta, satisfies:

$$N_F|0\rangle = 0 \quad \Rightarrow \quad \mathcal{H}_F|0\rangle = -\frac{\hbar\omega}{2}|0\rangle, \tag{22}$$

while the excited state $|1\rangle$, with one quantum, satisfies:

$$N_F|1\rangle = |1\rangle \quad \Rightarrow \quad \mathcal{H}_F|1\rangle = \frac{\hbar\omega}{2}|1\rangle. \tag{23}$$

The fermionic creation and annihilation operators act as follows:

$$a_F|0\rangle = 0, \tag{24}$$

$$a_F^\dagger|0\rangle = |1\rangle, \tag{25}$$

$$a_F^\dagger|1\rangle = 0. \tag{26}$$

Thus, the Hilbert space of a fermionic oscillator is two-dimensional. Notably, the ground state energy of a fermionic oscillator has the opposite sign compared to that of a bosonic oscillator. While the Hamiltonian for a bosonic oscillator can be expressed in terms of classical position and momentum variables, no direct classical analog exists for fermions. Constructing a classical Lagrangian for a fermionic oscillator requires anti-commuting classical variables, known as Grassmann variables, which are extensively studied in mathematics [18].

Finally, the fermionic oscillator, resembling a two-level system, has energy eigenvalues:

$$E_0 = -\frac{\hbar\omega}{2}, \quad E_1 = \frac{\hbar\omega}{2}. \tag{27}$$

4. Supersymmetric oscillator

A supersymmetric oscillator, serving as a foundational model in quantum field theory, represents the unification of a bosonic and a fermionic oscillator, both oscillating at a natural frequency ω [19, 20]. Conceptually, this system consists of one boson and one fermion confined within a harmonic potential, oscillating collectively at the same frequency ω. The bosonic component possesses an infinite ladder of energy eigenstates, whereas the fermionic component is restricted to two discrete energy eigenstates, corresponding to occupation numbers of $n_F = 0$ and $n_F = 1$. The supersymmetric oscillator is an elegant toy model that combines fundamental principles of quantum mechanics, thermodynamics, and symmetry into a unified framework. It provides insights into key features of supersymmetry, including vacuum energy cancelation, degeneracy in energy states, and the role of supercharges in the symmetry algebra. Its simplicity and mathematical richness make it an ideal testing ground for exploring theoretical predictions of supersymmetric quantum mechanics.

The Hamiltonian of the supersymmetric oscillator is expressed as:

$$\mathcal{H}_S = \mathcal{H}_B + \mathcal{H}_F = \hbar\omega\left(a_B^\dagger a_B + a_F^\dagger a_F\right), \tag{28}$$

where the subscripts S, B, and F denote supersymmetric, bosonic, and fermionic components, respectively. Using the number operators $N_B = a_B^\dagger a_B$ and $N_F = a_F^\dagger a_F$, the Hamiltonian can be re-expressed compactly as:

$$\mathcal{H}_S = \hbar\omega(N_B + N_F). \tag{29}$$

The energy eigenstates of the system are eigenstates of both N_B and N_F. Let us define these states as:

$$|N_B, N_F\rangle = |N_B\rangle \otimes |N_F\rangle, \tag{30}$$

where:

$$N_B|n_B\rangle = n_B|n_B\rangle, \quad n_B = 0,1,2,\ ... \tag{31}$$

$$N_F|n_F\rangle = n_F|n_F\rangle, \quad n_F = 0,1. \tag{32}$$

When the Hamiltonian acts on these eigenstates, the corresponding energy eigenvalues are:

$$\mathcal{H}_S|n_B, n_F\rangle = E_{n_B,n_F}|n_B, n_F\rangle = \hbar\omega(n_B + n_F)|n_B, n_F\rangle. \tag{33}$$

This leads to the energy spectrum:

$$E_n = \hbar\omega n, \quad n = n_B + n_F. \tag{34}$$

A distinctive feature of the supersymmetric oscillator is the absence of zero-point energy. Unlike a typical quantum harmonic oscillator, whose ground state has energy $E_0 = \hbar\omega/2$, the ground-state energy of the supersymmetric oscillator vanishes:

$$E_{0,0} = 0, \quad \text{where } a_B|0\rangle = 0,\ a_F|0\rangle = 0. \tag{35}$$

This vanishing of the ground-state energy is a hallmark of supersymmetry. The positive zero-point energy of the bosonic ground state, $E_B^0 = \hbar\omega/2$, is exactly canceled by the negative vacuum energy of the fermionic ground state, $E_F^0 = -\hbar\omega/2$. This phenomenon is not merely a coincidence but rather a direct consequence of the supersymmetry of the system. Canceling vacuum energy has profound implications in quantum field theory, particularly in addressing the cosmological constant problem.

Beyond the ground state, all energy eigenstates exhibit a two-fold degeneracy. For any bosonic quantum number n_B, there exist two states with identical energies: $|n_B, 1\rangle$ and $|n_B + 1, 0\rangle$. In other words, one state corresponds to $n_F = 0$ with $n_B = n$, while the other corresponds to $n_F = 1$ with $n_B = n - 1$. This degeneracy reflects the pairing of bosonic and fermionic states, a characteristic feature of supersymmetric systems. This pairing ensures that every energy level (except the ground state) contains a bosonic and a fermionic component. This symmetry plays a crucial role in the theoretical consistency of supersymmetric models, where the equal number of bosonic and fermionic degrees of freedom ensures the cancelation of quantum anomalies.

Supersymmetry is characterized by fermionic operators, known as supercharges, which generate transformations between bosonic and fermionic states. For the supersymmetric oscillator, the supercharges are defined as:

$$Q_1 = a_B^\dagger a_F, \quad Q_2 = a_F^\dagger a_B. \tag{36}$$

These supercharges commute with the Hamiltonian:

$$[Q_1, \mathcal{H}_S]_- = 0, \quad [Q_2, \mathcal{H}_S]_- = 0, \tag{37}$$

indicating that Q_1 and Q_2 are conserved quantities and act as symmetry generators. The supercharges transform bosonic states into fermionic states and vice versa while annihilating the ground state:

$$Q_1|0\rangle = 0, \quad Q_2|0\rangle = 0. \tag{38}$$

Therefore, Q_1 and Q_2 define conserved quantities of the system and would correspond to the generators of symmetries in the theory. Supercharges act on the paired states by transforming a bosonic state to a fermionic state and vice versa. The only exception is the ground state, which is annihilated by the supercharges. This is a typical situation for the supersymmetric system. There exist models in which one has a supersymmetry algebra, but the ground state has nonzero energy. In these models, within the language of quantum field theory, the ground state is not supersymmetric, and therefore, the model is not supersymmetric or in other words, supersymmetry is spontaneously broken. The anti-commutator of the supercharges gives the Hamiltonian:

$$[Q_1, Q_2]_+ = \frac{1}{\hbar\omega}\mathcal{H}_S, \tag{39}$$

forming a *graded Lie algebra*, which elegantly combines commutators and anti-commutators [21]. This algebra underpins the structure of supersymmetric quantum mechanics. Certainly, Eqs. (36), (37) and (39) indicate that, Q_1, Q_2 and $\mathcal{H}_S$ constitute an algebra that involves both commutators and anti-commutators. This algebra is the *graded Lie algebra* [21]. This algebra has a beautiful consequence, namely, if the ground state is invariant under the supersymmetry transformations (i.e., $Q_1|0\rangle = 0 = Q_2|0\rangle$ and $\langle 0|\mathcal{H}_S|0\rangle = \hbar\omega\langle 0|Q_1Q_2 + Q_2Q_1|0\rangle = 0$), then the ground state vanishes in supersymmetry theory. Q_2 is a hermitian conjugate of Q_1. We noticed that, since the Hamiltonian is a total square of a hermitian operator, all its states have nonnegative energies. The thermodynamic properties of the supersymmetric oscillator are derived from its canonical partition function. Given the energy eigenvalues $E_n = \hbar\omega n$, the partition function is:

$$\mathcal{Z} = \sum_{n_F=0}^{1}\sum_{n_B=0}^{\infty} e^{-\beta\hbar\omega(n_B+n_F)} = \frac{e^{\beta\hbar\omega}+1}{e^{\beta\hbar\omega}-1} = \coth\left(\frac{\beta\hbar\omega}{2}\right). \tag{40}$$

The partition function can calculate internal energy, entropy, and specific heat. The internal energy is:

$$\mathcal{U} = -\frac{\partial}{\partial\beta}\ln\mathcal{Z} = \frac{\hbar\omega}{\sinh(\beta\hbar\omega)}. \tag{41}$$

The entropy is:

$$\frac{S}{k_B} = \ln\mathcal{Z} - \beta\frac{\partial}{\partial\beta}\ln\mathcal{Z} = \ln\left[\coth\left(\frac{\beta\hbar\omega}{2}\right)\right] + (\beta\hbar\omega)\mathrm{cosech}(\beta\hbar\omega). \tag{42}$$

The specific heat is:

$$\frac{C_V}{k_B} = (\beta\hbar\omega)^2 \text{cosech}(\beta\hbar\omega)\coth(\beta\hbar\omega). \tag{43}$$

Alternatively, specific heat can be written as

$$\frac{C_V}{k_B} = 2(\beta\hbar\omega)^2 \frac{e^{\beta\hbar\omega}}{(e^{2\beta\hbar\omega}-1)}\left(\frac{e^{2\beta\hbar\omega}+1}{e^{2\beta\hbar\omega}-1}\right). \tag{44}$$

The behavior of the specific heat and the entropy of a supersymmetric oscillator concerning temperature is shown in **Figure 1**. As the temperature $T \to 0$, the specific heat of the supersymmetric oscillator $\frac{C_V}{k_B} \approx 2\left(\frac{\hbar\omega}{k_B T}\right)^2 e^{-\frac{\hbar\omega}{k_B T}}$. This shows that the specific heat falls exponentially to zero twice as fast as the simple quantum oscillator. On the other hand, at very high temperatures, the specific heat goes like $\left(1+\frac{\hbar\omega}{k_B T}\right)^2$, which is the square of the corresponding value of the heat capacity of a quantum oscillator at high temperatures. The anomalous peak at a certain temperature when the supersymmetric oscillator approaches low temperatures clearly shows the fermionic type of behavior. Therefore, the supersymmetric oscillator satisfies the third law of thermodynamics at low temperatures and satisfies the classical equipartition theorem at high temperatures. The entropy vanishes exponentially ($\approx\beta\hbar\omega e^{-\beta\hbar\omega}$) at very low temperatures and approaches infinity at very high temperatures. The supersymmetric oscillator exemplifies the deep interplay between quantum mechanics, thermodynamics, and symmetry. Its unique energy spectrum, vanishing ground-state energy, and paired degeneracy illustrate fundamental aspects of supersymmetry. Furthermore, its thermodynamic behavior provides insights into how supersymmetric systems bridge quantum and classical regimes. The model continues to be a cornerstone in theoretical physics, offering a platform for exploring more complex supersymmetric theories.

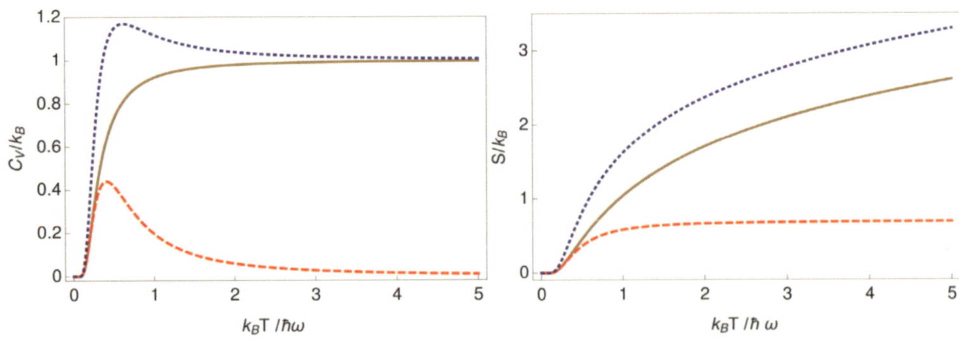

Figure 1.
The behavior of the (a) specific heat and (b) the entropy with respect to the dimensionless temperature $k_BT/\hbar\omega$, is plotted here. For a reference, the behaviors of the bosonic and fermionic oscillators are plotted. The thick curve represents the bosonic oscillator, the dashed one represents the fermionic oscillator, and the dotted one represents the supersymmetric oscillator. For a bosonic oscillator, the specific heat is given by the formula $C_V/k_B = (\beta\hbar\omega/2)^2 \text{Cosech}^2(\beta\hbar\omega/2)$ and that of a fermionic oscillator is $C_V/k_B = (\beta\hbar\omega/2)^2 \text{Sech}^2(\beta\hbar\omega/2)$. The entropy of a bosonic oscillator is given by $S/k_B = (\beta\hbar\omega/2)\coth(\beta\hbar\omega/2) - \ln[2\sinh(\beta\hbar\omega/2)]$ and that of a fermionic oscillator is given by $S/k_B = -(\beta\hbar\omega/2)\tanh(\beta\hbar\omega/2) + \ln[2\cosh(\beta\hbar\omega/2)]$.

5. Electron in a magnetic field: Example of supersymmetry in a real system

As a supplementary discussion, we now turn our attention to the celebrated problem of an electron in a magnetic field, commonly referred to as the "Landau problem" [7, 22]. This problem has become a cornerstone in condensed matter physics, primarily due to the pioneering contributions of Lev Landau, who laid the foundation for understanding the diamagnetism of free electron gases. The Landau problem is also essential for exploring quantum Hall effects, magnetic oscillations, and other phenomena involving charged particles in a magnetic field. Here, we investigate this problem's supersymmetric structure when the electron's spin angular momentum is incorporated into the analysis. The Hamiltonian for a nonrelativistic electron moving in a uniform magnetic field is expressed as:

$$\mathcal{H} = \frac{1}{2m}\left(\mathbf{P} - \frac{e}{c}\mathbf{A}\right)^2, \tag{45}$$

where m is the mass of the electron, $\mathbf{P}$ is the canonical momentum, e is the charge, c is the speed of light, and $\mathbf{A}$ is the magnetic vector potential associated with the external magnetic field $\mathbf{B} = \nabla \times \mathbf{A}$. For simplicity, we assume that the magnetic field is uniform and aligned along the z-axis ($\mathbf{B} = B\hat{z}$). Consequently, the electron's motion is confined to the xy-plane, while the magnetic field introduces a Lorentz force that causes the electron to undergo circular motion.

To simplify the problem, we define two operators:

$$a = \sqrt{\frac{c}{2e\hbar B}}(P_x + iP_y), \quad a^\dagger = \sqrt{\frac{c}{2e\hbar B}}(P_x - iP_y), \tag{46}$$

which satisfy the commutation relation:

$$\left[a, a^\dagger\right] = 1. \tag{47}$$

In terms of these ladder operators, the Hamiltonian can be written as:

$$\mathcal{H} = \hbar\omega_c\left(a^\dagger a + \frac{1}{2}\right), \tag{48}$$

where $\omega_c = \frac{eB}{mc}$ is the cyclotron frequency, representing the angular frequency of the electron's circular motion. The position and momentum operators are related to the ladder operators through:

$$P_x = p_x - \frac{eyB}{2c}, \quad P_y = p_y + \frac{exB}{2c}. \tag{49}$$

When the spin angular momentum of the electron is incorporated, the Hamiltonian gains an additional term representing the interaction between the spin and the magnetic field:

$$\mathcal{H} = \hbar\omega_c\left(a^\dagger a + \frac{1}{2}\right) - \frac{ge}{2mc}\vec{S}\cdot\vec{B}, \tag{50}$$

where $\vec{S}$ is the spin angular momentum operator, and g is the Landé g-factor. For a magnetic field aligned along the z-axis, $\vec{S} \cdot \vec{B} = S_z B$, and we use:

$$S_z = \frac{\hbar \sigma_z}{2}, \tag{51}$$

where σ_z is the Pauli spin matrix. The Hamiltonian then becomes:

$$\mathcal{H} = \hbar\omega_c \left(a^\dagger a + \frac{1}{2} - \frac{g}{4}\sigma_z \right). \tag{52}$$

For $g = 2$, the supersymmetric structure of the problem emerges clearly. The ground state with $\sigma_z = +1$ has precisely zero energy. The ground state with $\sigma_z = -1$ and the first excited state with $\sigma_z = +1$ become degenerate, forming paired states that are characteristic of supersymmetric systems. To explore this structure further, we introduce an operator:

$$\mathcal{Q} = \sqrt{\frac{1}{2m}} (\sigma_x P_x + \sigma_y P_y), \tag{53}$$

whose square is the Hamiltonian:

$$\mathcal{Q}^2 = \mathcal{H}. \tag{54}$$

The operator $\mathcal{Q}$ can be expressed in matrix form as:

$$\mathcal{Q} = \sqrt{\frac{1}{2m}} \begin{pmatrix} 0 & P_x - iP_y \\ P_x + iP_y & 0 \end{pmatrix} = \sqrt{\frac{\hbar\omega_c}{2}} \begin{pmatrix} 0 & a^\dagger \\ a & 0 \end{pmatrix}. \tag{55}$$

The operator $\mathcal{Q}$ acts as a ladder operator, raising the Landau level and the spin simultaneously or lowering both. Importantly, $\mathcal{Q}$ commutes with the Hamiltonian:

$$[\mathcal{Q}, \mathcal{H}] = 0, \tag{56}$$

indicating that $\mathcal{Q}$ does not alter the energy of the state. As a result, each Landau level exhibits degeneracy between the two spin states, forming pairs $|i\rangle$ and $\mathcal{Q}|i\rangle$. However, the lowest Landau level with spin-up is:

$$\begin{pmatrix} \psi_0^L \\ 0 \end{pmatrix}, \tag{57}$$

is annihilated by $\mathcal{Q}$ and remains nondegenerate. It also has precisely zero energy, consistent with supersymmetric systems:

$$\mathcal{H} = \mathcal{Q}^2 = 0. \tag{58}$$

To further elucidate the supersymmetric structure, we define fermionic operators:

$$\hat{f}^{\dagger} = \hat{\sigma}^{+} = \begin{pmatrix} 0 & 1 \\ 0 & 0 \end{pmatrix}, \quad \hat{f} = \hat{\sigma}^{-} = \begin{pmatrix} 0 & 0 \\ 1 & 0 \end{pmatrix}, \tag{59}$$

where $\hat{\sigma}^{\pm} = \frac{1}{2}(\sigma_x \pm i\sigma_y)$. Using these operators, the Hamiltonian in Eq. (52) can be rewritten as:

$$\mathcal{H} = \hbar\omega_c\left(a^{\dagger}a + \frac{1}{2}\right) + \hbar\omega_c\left(\hat{f}^{\dagger}\hat{f} - \frac{1}{2}\right), \tag{60}$$

or equivalently:

$$\mathcal{H} = \hbar\omega_c\left(a^{\dagger}a + \hat{f}^{\dagger}\hat{f}\right). \tag{61}$$

This reformulation highlights the resemblance of the problem to a supersymmetric oscillator, where the Hamiltonian is a combination of bosonic and fermionic contributions.

The thermodynamic properties of the system are determined from the partition function. For an electron in a homogeneous magnetic field, the partition function is given by:

$$\mathcal{Z} = \mathcal{G}\frac{1}{2}\text{Cosech}\left(\frac{\beta\hbar\omega_c}{2}\right), \tag{62}$$

where:

$$\mathcal{G} = \left(\frac{m\omega_c\mathcal{A}}{2\pi\hbar}\right), \tag{63}$$

is the degeneracy factor, and $\mathcal{A}$ is the area in the xy-plane. This term does not depend on temperature and thus does not directly affect thermodynamic properties. When spin is included, the partition function becomes:

$$\mathcal{Z} = \mathcal{G}\coth\left(\frac{\beta\hbar\omega_c}{2}\right). \tag{64}$$

This partition function resembles that of a supersymmetric oscillator, with ω replaced by ω_c. For example, the specific heat is:

$$C_V = k_B\left(\frac{\beta\hbar\omega_c}{2}\right)^2 \text{Cosech}^2\left(\frac{\beta\hbar\omega_c}{2}\right). \tag{65}$$

Figure 2 illustrates the specific heat behavior for an electron in a magnetic field, both with and without spin. These results demonstrate the supersymmetric nature of the system when spin is taken into account. The analogy with the supersymmetric oscillator further solidifies the profound connection between quantum mechanics and supersymmetry.

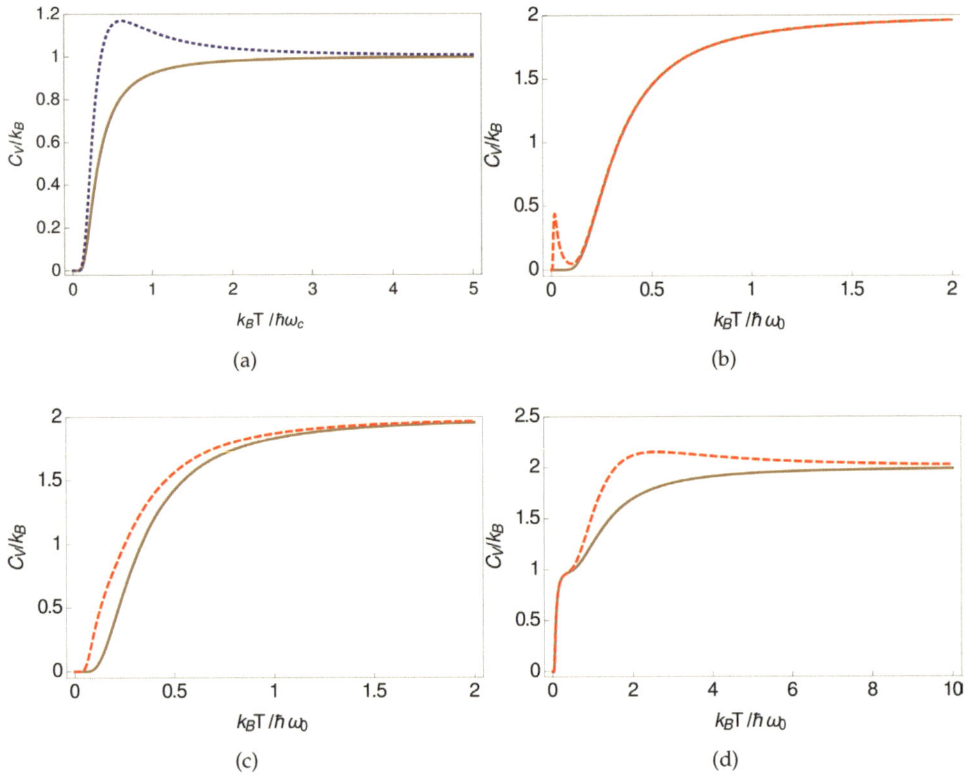

Figure 2.
(a) The behavior of the specific heat of an electron in a magnetic field with respect to the dimensionless temperature $k_B T/\hbar\omega_c$ *is plotted. The dotted curve includes spin, and the solid curve omits spin. (b, c, d) The specific heat for different cyclotron frequencies* $\omega_c = 0.04\omega_0$, $0.4\omega_0$, *and* $4\omega_0$, *respectively. The thick curve represents the case without spin, and the dashed curve includes spin.*

6. Examples of supersymmetry breaking

6.1 Introduction of a confinement to the Landau model

With the introduction of a confinement frequency into the problem, we observe a profound change in the behavior of the energy spectrum of an electron in a magnetic field. Specifically, the ground state energy of the electron can no longer be zero under any chosen values of the Landé g-factor or the spin projection σ_z. This arises naturally when a symmetric and isotropic two-dimensional parabolic confinement is introduced, described by the potential:

$$V(x,y) = \frac{1}{2} m\omega_0^2 (x^2 + y^2), \tag{66}$$

where m is the mass of the electron, and ω_0 is the confinement frequency. This harmonic potential confines the electron within the xy-plane and modifies the quantized energy levels induced by the magnetic field. Such a model is physically relevant in systems like quantum dots, where the electron motion is confined to a plane and subjected to an external magnetic field.

The total Hamiltonian of the system, incorporating both the magnetic field and the parabolic confinement, is given by:

$$\mathcal{H} = \frac{1}{2m}\left[\left(p_x - \frac{eyB}{2c}\right)^2 + \left(p_y + \frac{exB}{2c}\right)^2\right] + \frac{1}{2}m\omega_0^2(x^2 + y^2), \tag{67}$$

The first term accounts for the electron's kinetic energy in the presence of a Lorentz force, and the second term represents the parabolic trapping potential. This Hamiltonian is widely known in the literature as the *Fock-Darwin model* [23, 24]. The eigenvalues of the Hamiltonian in Eq. (67) lead to the energy spectrum:

$$\varepsilon_{n_1,n_2} = \hbar\sqrt{\omega_0^2 + \left(\frac{\omega_c}{2}\right)^2}(n_1 + n_2 + 1) + \frac{1}{2}\hbar\omega_c(n_1 - n_2), \tag{68}$$

where $\omega_c = \frac{eB}{mc}$ is the cyclotron frequency, and n_1, n_2 are quantum numbers associated with the motion in the xy-plane. This energy spectrum exhibits several unique and important features:

- *Two-dimensional harmonic oscillator*: In the absence of the magnetic field ($\omega_c = 0$), the energy levels reduce to those of a two-dimensional isotropic harmonic oscillator:

$$\varepsilon_{n_1,n_2} = \hbar\omega_0(n_1 + n_2 + 1).$$

- *Landau levels*: When the parabolic confinement is absent ($\omega_0 = 0$), the energy levels correspond to the Landau levels of a free particle in a magnetic field:

$$\varepsilon_{n_1,n_2} = \hbar\omega_c\left(n_1 + \frac{1}{2}\right).$$

- *Transition regime*: For intermediate values of ω_0 and ω_c, the energy spectrum represents a hybrid of the harmonic oscillator and Landau quantization. This regime is particularly interesting for systems like quantum dots, where both magnetic quantization and confinement play a significant role.

To analyze the Hamiltonian further, we introduce the quantum mechanical operators [25]:

$$a_{\pm} = \frac{1}{2l}[x \mp iy] - \frac{l}{2}\left[\partial_x \mp i\partial_y\right], \tag{69}$$

where $l = \sqrt{\frac{\hbar}{m\omega}}$ is the Fock-Darwin length, a characteristic length scale determined by the effective frequency $\omega = \sqrt{\omega_0^2 + \frac{\omega_c^2}{4}}$. These ladder operators simplify the analysis of the Hamiltonian, which can be rewritten as:

$$\mathcal{H}_0 = \left(a_+^\dagger a_+ + \frac{1}{2}\right)\hbar\Omega_+ + \left(a_-^\dagger a_- + \frac{1}{2}\right)\hbar\Omega_-, \tag{70}$$

where the effective frequencies Ω_+ and Ω_- are defined as:

$$\Omega_+ = \Omega + \frac{\omega_c}{2}, \quad \Omega_- = \Omega - \frac{\omega_c}{2}, \tag{71}$$

and $\Omega = \sqrt{\omega_0^2 + \frac{\omega_c^2}{4}}$.

The energy eigenvalues are now expressed as:

$$\varepsilon_{n_+,n_-} = \left(n_+ + \frac{1}{2}\right)\hbar\Omega_+ + \left(n_- + \frac{1}{2}\right)\hbar\Omega_-, \tag{72}$$

where $n_+, n_- = 0,1,2, \ldots$. This representation clearly shows the splitting of energy levels due to the combined effects of confinement and the magnetic field.

From Eq. (74), it is evident that the ground state energy of the system is:

$$\varepsilon_{0,0} = \frac{\hbar}{2}(\Omega_+ + \Omega_-). \tag{73}$$

Unlike the case of a pure Landau quantization, where the ground state energy can vanish for specific spin configurations ($g = 2, \sigma_z = +1$), the presence of the parabolic confinement ensures that the ground state energy is always nonzero. In the limit of a strong magnetic field ($\omega_c \gg \omega_0$), the confinement potential becomes negligible, and the energy levels approach the Landau levels:

$$\varepsilon_{n_+,n_-} \approx \hbar\omega_c\left(n_+ + \frac{1}{2}\right). \tag{74}$$

This limit highlights the transition between confinement-dominated and magnetic field-dominated regimes.

The total partition function of the system can be expressed as:

$$\mathcal{Z} = \frac{1}{2}\operatorname{csch}\left(\frac{\beta\hbar\Omega_+}{2}\right)\operatorname{csch}\left(\frac{\beta\hbar\Omega_-}{2}\right)\cosh\left(\frac{\beta\hbar\omega_c}{2}\right), \tag{75}$$

where $\beta = (k_B T)^{-1}$ is the inverse temperature. This expression incorporates contributions from both the orbital motion (*via* Ω_+ and Ω_-) and the spin degree of freedom (*via* ω_c). The thermodynamic quantities, such as internal energy, entropy, and specific heat, can be directly derived from the partition function. For example, the specific heat is given by:

$$C_V = k_B\beta^2 \frac{\partial^2}{\partial\beta^2}\ln\mathcal{Z}. \tag{76}$$

The Fock-Darwin model provides a robust framework for understanding the behavior of electrons in two-dimensional systems under the simultaneous influence of magnetic fields and confinement potentials. Its energy spectrum bridges two limiting cases: the two-dimensional harmonic oscillator and the Landau quantization. Including a parabolic confinement introduces a nonzero ground state energy, which fundamentally alters the thermodynamic properties of the system. This model finds direct applications in the study of quantum dots, where the interplay between confinement and magnetic quantization determines the electronic properties. It also serves as a testing ground for understanding mesoscopic systems' transition between classical and quantum regimes. Furthermore, the parallels between the Fock-Darwin model and supersymmetric oscillators underscore the deep connections between symmetry and the physical properties of quantum systems.

6.2 Jaynes-Cumming model: Mixture of bosonic modes and spin

Another fascinating example of supersymmetry breaking arises when we consider a system where a two-level system (a "fermionic oscillator") is coupled to a bosonic oscillator. This configuration resembles the supersymmetric oscillator because it incorporates a two-level system, analogous to a spin operator, and an oscillator exhibiting bosonic modes. Such a system is commonly called the *Jaynes-Cummings model* [26]. This model plays a pivotal role in quantum optics and cavity quantum electrodynamics, describing the interaction between light (bosonic modes) and matter (two-level atomic systems). The Hamiltonian for the Jaynes-Cummings (JC) model is expressed as:

$$H_{JC} = \hbar\omega\left(a^\dagger a + \frac{1}{2}\right) + \frac{1}{2}\varepsilon\sigma_z + \lambda\left(a\sigma_+ + a^\dagger\sigma_-\right), \tag{77}$$

where:

- $a^\dagger$ and a are the creation and annihilation operators for the bosonic field with frequency ω,
- σ_z, σ_+, and σ_- are the Pauli operators describing the two-level fermionic system,
- ε is the energy spacing of the two-level system,
- λ is the coupling strength between the bosonic and fermionic modes, and
- $\sigma_\pm = \frac{1}{2}\left(\sigma_x \pm i\sigma_y\right)$ are the raising and lowering operators for the two-level system.

The first term in H_{JC} represents the energy of the bosonic oscillator, the second term corresponds to the energy of the two-level system, and the third term captures the interaction between the bosonic field and the two-level fermionic system. The energy spectrum of the Jaynes-Cummings model is given by:

$$E_n^\pm = \left(n + \frac{1}{2}\right)\hbar\omega \pm \frac{\Delta_n}{2}, \tag{78}$$

where:

$$\Delta_n = \sqrt{\delta^2 + 4\lambda^2(n+1)}. \tag{79}$$

Here, $\delta = \omega - \varepsilon$ is the detuning factor, which measures the mismatch between the bosonic field frequency ω and the two-level system's energy splitting ε.

Eq. (78) reveals that the interaction between the bosonic and fermionic components results in a splitting of the energy levels. This coupling breaks the supersymmetry of the system in a significant way:

- *Ground state energy*: Unlike the supersymmetric oscillator, where the ground state energy is precisely zero, the ground state energy of the Jaynes-Cummings model is nonzero due to the coupling λ.

- *Energy splitting*: The coupling introduces a term proportional to Δ_n, which modifies the degeneracy of the energy levels and disrupts the balanced pairing of bosonic and fermionic states.

Thus, the interaction between the spin (fermionic mode) and the bosonic oscillator explicitly breaks the supersymmetry of the system. The Ja ynes-Cummings model provides a quantum mechanical description of light-matter interactions. It describes how a quantized electromagnetic field (bosonic mode) interacts with a two-level atom (fermionic mode). This coupling lies at the heart of phenomena such as Rabi oscillations, vacuum Rabi splitting, and quantum state transfer between light and matter.

Interestingly, the model suggests an inherent *Bose-Fermi duality* as it combines bosons in light quanta and fermions in the form of two-level systems. This duality hints at a hidden supersymmetry in the system. However, as shown above, the supersymmetry is spontaneously broken by the interaction term involving λ. Although supersymmetry is broken in the generic Jaynes-Cummings model, it can be restored under specific conditions. Studies have shown that supersymmetry can re-emerge when the model parameters satisfy certain integer relationships [27, 28]. For example:

- *Special tuning of parameters*: If the coupling strength λ, detuning δ, and bosonic frequency ω are fine-tuned to satisfy specific ratios, the degeneracies in the energy spectrum may be restored.

- *Symmetry in the interaction*: Supersymmetry is restored when the interaction term aligns symmetrically with the intrinsic energy scales of the bosonic and fermionic subsystems.

These findings highlight the interplay between symmetry breaking and restoration in coupled quantum systems, demonstrating the delicate balance required to preserve supersymmetric structures. The Jaynes-Cummings model provides a striking example of supersymmetry breaking in a physical system. The model describes light-matter interactions at a fundamental level by coupling a two-level fermionic system to a bosonic mode. The nonzero ground state energy and modified energy spectrum underscore the spontaneous breaking of supersymmetry. However, the hidden Bose-Fermi duality and the possibility of restoring supersymmetry through specific parameter tuning reveal the deep connections between quantum optics, supersymmetry, and the fundamental description of light-matter interactions. This model, therefore, serves as a vital bridge between theoretical concepts in supersymmetric quantum mechanics and practical applications in quantum information, quantum computing, and cavity quantum electrodynamics. Its rich structure inspires advancements in understanding quantum systems at the interface of light and matter.

7. Conclusion

In this chapter, we have conducted an extensive study of the thermodynamic properties of the supersymmetric oscillator, exploring its behavior in various temperature regimes and its connections to supersymmetric quantum mechanics. The results of this study provide deep insights into the role of supersymmetry in physical systems and its influence on thermodynamic quantities. One of the key findings is the specific heat behavior of the supersymmetric oscillator. At very low temperatures, the specific heat vanishes exponentially, consistent with the third law of thermodynamics. This rapid decay highlights the suppression of accessible microstates as the temperature approaches absolute zero, a hallmark of quantum systems with well-separated energy levels. At high temperatures, the specific heat asymptotically approaches its classical equipartition value, confirming the crossover from quantum to classical behavior. Notably, the supersymmetric oscillator exhibits an anomalous peak in its specific heat at intermediate temperatures, similar to a two-level or single spin system. This feature reflects the fermionic component's contribution to the oscillator's thermodynamics, emphasizing the interplay between the bosonic and fermionic degrees of freedom inherent in supersymmetric systems.

Beyond the supersymmetric oscillator, we have explored the problem of an electron in a magnetic field, commonly known as the Landau problem. When the electron's spin angular momentum is considered, the system naturally acquires a supersymmetric structure. This surprising result underscores the fundamental nature of supersymmetry in quantum systems with balanced fermionic and bosonic contributions. However, including a confinement frequency, such as that in the Fock-Darwin model, breaks the supersymmetry of the system. This breaking arises from introducing an additional potential that lifts the degeneracy between the bosonic and fermionic contributions, leading to a nonzero ground state energy and a modified energy spectrum. We have also examined the Jaynes-Cummings model, a paradigmatic system in quantum optics that describes the interaction between a two-level atomic system and a quantized bosonic field. The Jaynes-Cummings model resembles the supersymmetric oscillator, incorporating both bosonic and fermionic degrees of freedom. However, our analysis shows that supersymmetry does not exist in this model under generic conditions. The Jaynes-Cummings model's interaction between the bosonic and fermionic components introduces an energy splitting that breaks the pairing of bosonic and fermionic states. Consequently, the ground state energy of the Jaynes-Cummings system is nonzero, indicating spontaneous breaking of supersymmetry. Interestingly, supersymmetry can be restored under specific tuning of the model parameters, where the degeneracies of the energy spectrum are recovered.

Through these studies, we have demonstrated that while elegant and powerful, supersymmetry is sensitive to the details of system interactions and external conditions. The spontaneous breaking of supersymmetry observed in the Landau problem with confinement and the Jaynes-Cummings model highlights the intricate balance required to maintain this symmetry. These findings underscore the broader implications of supersymmetry in quantum mechanics, statistical mechanics, and quantum field theory. The results presented in this paper provide a unified perspective on the thermodynamic and quantum properties of supersymmetric systems. They also reveal the connections between supersymmetric oscillators, electrons in magnetic fields, and light-matter interactions in quantum optics. These insights could be a foundation for further research into applying supersymmetric principles in mesoscopic systems, quantum technologies, and beyond. Future investigations could explore the restoration of supersymmetry in broken systems and the potential realization of supersymmetric quantum devices, opening new avenues for integrating symmetry concepts into experimental and applied physics.

Author details

Jishad Kumar
Department of Applied Physics, Aalto University, Espoo, Finland

*Address all correspondence to: kumar.jishad@gmail.com

References

[1] Muller I. A History of Thermodynamics: The Doctrine of Energy and Entropy. Heidelberg, Berlin: Springer-Verlag; 2007. pp. 13-29

[2] Dugdale JS. Entropy and its Physical Meaning. London: Taylor and Francis; 1996. pp. 39-55

[3] Klein MJ. Maxwell, his demon, and the second law of thermodynamics. American Scientist. 1970;**58**:84-97

[4] Planck M. On the law of distribution of energy in the normal spectrum. Annalen der Physik. 1901;**4**:553-562. DOI: 10.1002/andp.19013090310

[5] Gallavotti G, Reiter WL, Yngvason J, editors. Boltzmann's Legacy: Lectures in Mathematics and Physics. Zürich: European Mathematical Society Publishing House; 2008. pp. 99-103

[6] Gibbs JW. The Scientific Papers of J. Willard Gibbs, Vol. 1 Thermodynamics. USA: Dover; 1961

[7] Landau L. Das Dämpfungsproblem in der Wellenmechanik. Zeitschrift für Physik. 1927;**45**:430-441. DOI: 10.1007/BF01343064

[8] von Neumann J. Thermodynamik quantenmechanischer Gesamtheiten. In: Nachrichten von der Gesellschaft der Wissenschaften zu Göttingen, Mathematisch-Physikalische Klasse. 1927. pp. 273-291. Available from: http://eudml.org/doc/59231

[9] Dirac PAM. The basis of statistical quantum mechanics. Mathematical Proceedings of the Cambridge Philosophical Society. 1929;**25**(1):62-66. DOI: 10.1017/S0305004100018570

[10] Dirac PAM. On the annihilation of electrons and protons. Mathematical Proceedings of the Cambridge Philosophical Society. 1930;**26**(3):361-375. DOI: 10.1017/S0305004100016091

[11] Dirac PAM. Note on the interpretation of the density matrix in the many-electron problem. Mathematical Proceedings of the Cambridge Philosophical Society. 1931; **27**(2):240-243

[12] Callen HB. Thermodynamics and an Introduction to Thermostatistics. Singapore: John Wiley; 1985

[13] Witten E. Dynamical breaking of supersymmetry. Nuclear Physics B. 1981;**188**:513-554. DOI: 10.1016/0550-3213(81)90006-7

[14] Cooper F, Khare A, Sukhatme U. Supersymmetry and quntum mechanics. Physics Reports. 1995;**251**(5-6):267-285. DOI: 10.1016/0370-1573(94)00080-M

[15] Bagger J, Wess J. Supersymmetry and Supergravity. Princeton: Princeton University Press; 1992. ISBN: 9780691025308

[16] Zettili N. Quantum Mechanics: Concepts and Applications. 2nd ed. Chichester: John-Wiley; 2009. ISBN: 978-0-470-74656-1

[17] Das A. Field Theory: A Path Integral Approach. Singapore: World Scientific Publishing; 1993

[18] Kleinert H. Path Integrals in Quantum Mechanics, Statistics. Polymer Physics and Financial Markets. Singapore: World Scientific; 2004

[19] Müller-Kirsten HJW, Wiedmann A. Introduction to Supersymmetry. 2nd ed. Singapore: World Scientific; 2010

[20] Martin SP. A supersymmetry primer. arxiv: hep-ph/9709356. 1997 [Accessed: January 27, 2016]

[21] Corwin L, Ne'eman Y, Sternberg S. Graded Lie algebras in mathematics and physics (Bose-Fermi symmetry). Reviews of Modern Physics. 1975;**47**:573

[22] Landau LD. Diamagnetismus der Metalle. Zeitschrift für Physik. 1930;**64**:629

[23] Fock V. Bemerkung zur Quantelung des harmonischen Oszillators im Magnetfeld. Zeitschrift für Physik. 1928; **48**:446-448. DOI: 10.1007/BF01390750

[24] Darwin CG. The diamagnetism of the free electron. Mathematical Proceedings of the Cambridge Philosophical Society. 1931;**27**(1):86-90. DOI: 10.1017/S0305004100009373

[25] Rössler U. In: Kramer B, editor. Quantum Coherence in Mesoscopic Systems. New York: Plenum Press; 1991. pp. 45-62

[26] Jaynes ET, Cummings FW. Comparison of quantum and semiclassical radiation theories with application to the beam maser. Proceedings of the IEEE. 1963;**51**:89-109. DOI: 10.1109/PROC.1963.1664

[27] Andreev VA, Lerner PB. Supersymmetry in the Jaynes-Cummings model. Physics Letters A. 1989;**134**(8-9):507-511

[28] Castanos O. Supersymmetry in the Jaynes-Cummings model. AIP Conference Proceedings. 2013;**1540**:77

Chapter 5

Black Hole as an Information Mirror

Xuanhua Wang

Abstract

Black holes, as described within the framework of effective field theory, radiate particles in thermal spectra, with temperatures proportional to the surface gravities on the event horizons. If a black hole forms from collapsing matter in a pure state, Hawking's calculation suggests that the resulting radiation is in a mixed state by the end of the evaporation process, violating the unitarity of quantum mechanics. This contradiction led to a long-standing puzzle regarding information conservation in black holes. After the discovery of the AdS/CFT correspondence, it became more accepted that black hole dynamics should be unitary, even though concrete calculations were still lacking. Recent breakthroughs, particularly computations of radiation entropy using quantum extremal surfaces, have successfully demonstrated the conservation of von Neumann entropy during black hole evaporation, while quantum information theory has enabled the development of protocols for decoding information from Hawking radiation. However, challenges persist in understanding the precise mechanisms by which information from infalling matter becomes encoded in radiation and how external observers might retrieve it. Nevertheless, recent advances suggest these questions are becoming increasingly tractable, offering valuable insights into the conditions and processes that enable information recovery from black hole radiation.

Keywords: black hole information paradox, Hawking radiation, entanglement entropy, unitarity, information decoding

1. Introduction

Not long ago, black holes were often pictured as cosmic vacuums with singularities at their core—giant, dark objects that devour everything crossing their event horizons, allowing no escape or return. However, this perspective has undergone a profound transformation: black holes are no longer viewed solely as information destroyers but, intriguingly, as mirrors that can reflect information back into the universe in a changed form. In this chapter, we will delve into the fascinating and evolving problem of black hole information.

The study of black holes is not only essential for cosmology but also instrumental for quantum gravity. Black holes provide an ideal platform to explore the intersection of general relativity and quantum mechanics [1–4]. The black hole information paradox serves as a prime example of the apparent conflict between these two foundational frameworks. Proposed by Stephen Hawking in the 1970s, the paradox stems from his groundbreaking discovery that black holes emit thermal radiation—now known as Hawking radiation—with a temperature determined by the event horizon's surface gravity [5–7]. This discovery implied that black holes, once thought to trap everything forever, could evaporate over time. If the matter that collapses into a black hole begins in a pure quantum state, the emitted thermal radiation seemingly carries no information about the initial state. By the time the black hole fully evaporates, the remaining mixed state would violate the principle of unitarity, which asserts that quantum information must always be conserved. However, allowing information to escape will require modification of the entanglement between the emitted radiation and the black hole, which will cause a "firewall" on the horizon, violating equivalence principle [8, 9]. This paradox raised one of the most profound and challenging questions in modern theoretical physics [10–12].

One critical aspect of the black hole information paradox is around the von Neumann entropies of a black hole and its Hawking radiation. In an evaporating black hole, the emitted radiation remains entangled with the black hole itself, leading to an increase in the radiation's von Neumann entropy. This behavior seems to challenge the Bekenstein bound, which states that the total entropy of a system cannot exceed the entropy proportional to its boundary area [13–15]. Yet, calculations from effective field theory suggest the violation of this bound due to the monotonically increasing entropy in Hawking radiation. In 1993, Don Page introduced the Page curve, a theoretical prediction of how entropy evolves during black hole evaporation [16, 17]. Initially, entropy rises as radiation is emitted. However, if information is preserved, the entropy must eventually decline as the black hole nears complete evaporation, revealing the encoded information. This characteristic behavior of entropy became a central feature in understanding the paradox.

The debate on whether black holes conserve information gained public attention with the famous Hawking-Preskill-Thorne bet [18]. Stephen Hawking and Kip Thorne bet John Preskill that information was irrevocably lost in black holes. In 2004, after significant theoretical developments, including insights from the AdS/CFT correspondence, Hawking conceded the bet, acknowledging that information is likely preserved during black hole evaporation.

A significant breakthrough emerged in recent years with the discovery of a generalized gravitational entropy formula, which is also referred to as the "Island formula" [19–21]. The formula introduces isolated regions—or "islands"—inside the black hole that contribute to the entanglement entropy at late stages of evaporation. Calculations based on this formula have offered a robust demonstration of entropy conservation in gravitational systems [22–27].

Closely tied to entropy conservation is the question of when information can be retrieved from black hole radiation. While logically this question assumes entropy conservation in black holes, it was addressed earlier than the Island formula through arguments based on quantum information theory. Analyses suggest that after the Page time, information can be extracted from Hawking radiation at close to the maximal speed [28–30]. This finding effectively transforms a black hole from an information destroyer into an information mirror. Rapid advancements in quantum information science and quantum computing over the last decade have significantly influenced our understanding of fundamental physics. These developments also sparked much intellectual curiosity about whether we could realistically decode information from Hawking radiation if a black hole was created in a laboratory and all its radiations were carefully collected. Consequently, various decoding programs were investigated [31–38]. These studies using quantum information tools have provided tremendous new insights than previous anticipated and extended the horizon of our understandings.

This chapter explores the historical context, key theoretical breakthroughs, and ongoing efforts to resolve the black hole information paradox. In Section 2, we discuss the calculation of the von Neumann entropy for black holes, along with the Island formula. Section 3 focuses on how quantum information theory can help address two crucial questions about black hole information: When does information escape the black hole? And how can we decode this information from the emitted radiation? These discussions will shed light on the path toward a deeper understanding of black hole information dynamics.

2. Entanglement entropy of black holes

While the second law of thermodynamics suggests a perpetual increase in the thermal entropy of any closed system, the von Neumann entropy of a quantum system is conserved under unitary evolution. This principle directly conflicts with Hawking's radiation formula derived from effective field theory. For massive black holes, where effective field theory is expected to remain valid even near the event horizon, this inconsistency became a focal point of research. Therefore, the computation of entropy is essential for the understanding of the problem. In this section, we introduce the development in calculating the von Neumann entropy for gravitational systems.

2.1 Ryu-Takayanagi formula and quantum extremal surface

The von Neumann entropy can be viewed as a quantum extension of the Shannon entropy, which measures the uncertainty associated with the states in a system or the amount of information one obtains once the state is known. For quantum systems, the uncertainty arises from the entanglement with the outside. Therefore, the von Neumann entropy measures the amount of entanglement a system has with external systems [39]. The definition of von Neumann entropy is given by

$$S(\rho) = -\mathrm{tr}\rho \log \rho, \tag{1}$$

where ρ is the density operator of the system. For a pure system AB consists of A and its complement $B = \overline{A}$, the entropy satisfies $S(\rho_A) = S(\rho_B)$ and represents the entanglement between the two systems. Since we usually consider black holes formed by pure-state matter for simplicity, the entropy of a black hole is also the entropy of its radiation. Notably, to compute the von Neumann entropy, one has to know the quantum state of the system beforehand. However, it is not easy to do in case of black holes due to the lack of well-established theory of quantum gravity. The effective theory calculation given by Hawking yields a thermal density matrix for the radiation:

$$\rho \propto \bigotimes_{\omega, l, m} \left(\sum_n |n\rangle\langle n|_{\omega,l,m} P(\omega, l) e^{-\beta\omega n} \right), \tag{2}$$

where $P(\omega, l)$ is the gray-body factor and n is the particle number [5–7]. Noticeably, the low-energy modes are nearly maximally entangled. This density matrix does not respect unitarity in the later stages of a black hole's evolution due to monotone increase of entropy.

In conformal field theories, Ryu and Takayanagi proposed a new approach to compute entropy without relying directly on the exact density matrix [40, 41]. They demonstrated that the entanglement entropy of a region in the boundary CFT corresponds to the minimal area of a homologous surface in the bulk, a relationship known as the Ryu-Takayanagi (RT) formula. The formula was explicitly calculated in specific examples but conjectured as a more universal formula for the subsystem entanglement entropy S_A of CFT_{d+1} given by

$$S_A = \frac{\mathrm{Area}(\gamma_A)}{4G_N^{d+2}}, \tag{3}$$

where the manifold γ_A is the d-dimensional minimal static surface with the boundary ∂A. This surface is also referred to as the RT surface (see **Figure 1**). This conjecture offers a geometric method for calculating entanglement entropy, bypassing the explicit use of the exact density matrix. Instead, the entropy is determined purely from the geometry. This key insight has played a significant role in addressing the black hole information paradox and deepened our understanding of the relationship between geometry and quantum information.

For quantum fields in gravitational backgrounds, such as those involving black holes, quantum corrections to the RT formula introduce an additional term corresponding to the bulk entanglement entropy [19, 21]. This modified formula has been derived through various approaches, leading to an extension of the original RT prescription. Throughout our discussion, we will assume that a black hole and its radiation together form a pure state. In this case, the entropies of the black hole and its Hawking radiation are identical, and we use these terms interchangeably. The modified entropy formula is given by [19–21, 41–43].

$$S_{\mathrm{vN}} = \min_X \left\{ \mathrm{ext}_X \left[\frac{\mathrm{Area}(X)}{4G_N} + S_{\mathrm{semi-cl}(\Sigma_X)} \right] \right\}, \tag{4}$$

where X is the codimension-2 surface, Σ_X is the region between the boundary of the region of interest and X, and $S_{\mathrm{semi-cl}(\Sigma_X)}$ is the entanglement entropy of quantum fields on Σ_X in the semi-classical description where gravity is treated classically. The cutoff surface marks the boundary beyond which gravitational effects are considered negligible, and the spacetime geometry is treated as fixed. A demonstration of the surfaces in an evaporating black hole is shown in **Figure 2**. The surface that extremizes the generalized entropy, defined by the expression inside the brackets of Eq. (4), is termed the quantum extremal surface. The quantum extremal surface can be viewed as an extension of the RT surface which includes corrections from quantum gravity.

This entropy formula is also referred to as the "island formula" in literature, particularly in the context of computing the entropy of Hawking radiation. When computing the radiation entropy, the formula takes the form:

$$S_{\mathrm{vN}}(\mathrm{Rad}) = \min_X \left\{ \mathrm{ext}_X \left[\frac{\mathrm{Area}(X)}{4G_N} + S_{\mathrm{semi-cl}}(\Sigma_{\mathrm{Rad}} \cup \Sigma_{\mathrm{Island}}) \right] \right\}, \tag{5}$$

where Σ_{Rad} is the region outside cutoff surface and Σ_{Island} is the region inside the quantum extremal surface. Notably, when computing the semi-classical entropy,

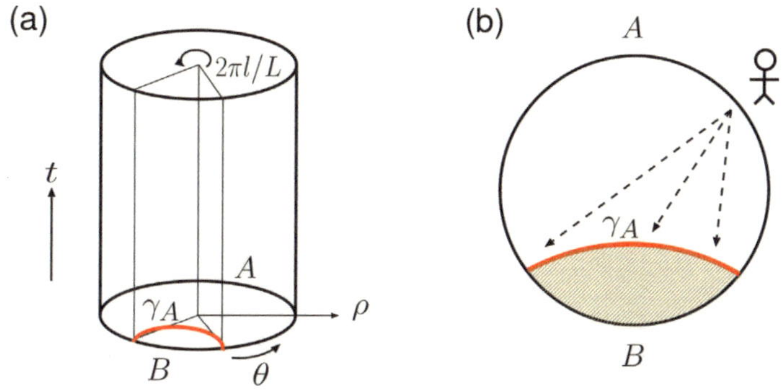

Figure 1.
An diagrammatic illustration of an RT surface in AdS_3/CFT_2. (a) AdS_3 space in the global coordinate and CFT_2 living on its boundary. γ_A in red is the RT surface for the subsystem A. (b) The RT surface on a time slice is the geodesic in AdS_3, and it can be viewed as a holographic screen of the subsystem B. Figure from Ref. [41].

contributions from both the region inside the quantum extremal surface—the island region—and the region outside the cutoff surface must be included.

2.2 Derivation of the island formula

The most direct way to derive the formula is through the gravitational path integral and the replica trick. Here, we present a brief derivation of the gravitational entropy formula. Interested readers can refer to Refs. [19, 23, 44] for detailed discussions. In quantum field theory, the entanglement entropy can be calculated using the replica trick. Suppose we have n copies of the system, the entropy for a subsystem A can then be written as [44].

$$S_A = -\mathrm{tr}_A \rho_A \log \rho_A = -\lim_{n\to 1} \partial_n \log \mathrm{tr}_A \rho_A^n. \tag{6}$$

Here, $\mathrm{tr}_A \rho_A^n$ can be computed on the n-fold cover of the original spacetime, where replicas of the original spacetime sheet glued together with a cut along A. This construction is illustrated in **Figure 3**, where fields marked by the same colors are identified. From the figure, we can directly read

$$\mathrm{tr}_A\left(\rho_A^n\right) = \frac{Z_n}{Z^n}, \tag{7}$$

where Z_n is the n-sheet partition function. From Eq. (6), the entropy for quantum fields in region A can be expressed as

$$S_A = -\lim_{n\to 1} \partial_n (\log Z_n - n \log Z). \tag{8}$$

For conformal fields living on general manifolds with nontrivial gravity or topology, the total partition function includes both the gravitational contribution and the partition function for the quantum fields:

$$Z = \int Dg e^{-S_E[g]} Z_{CFT}, \tag{9}$$

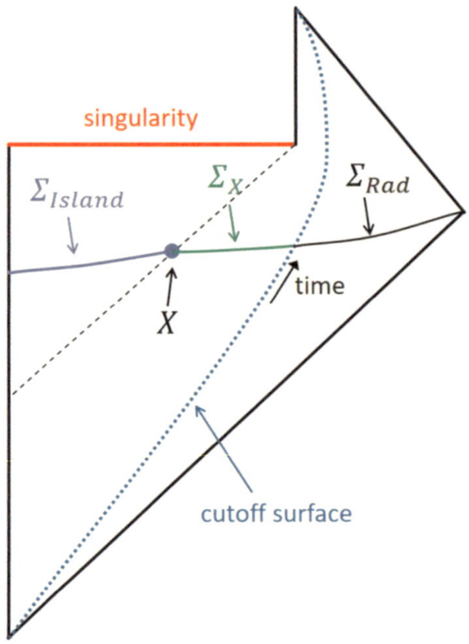

Figure 2.
Quantum extremal surface and island in the Penrose diagram for an evaporating black hole after the Page time. To compute the entropy of Hawking radiation outside the cutoff surface, one computes the surface that extermizes the generalized entropy inside brackets of Eq. (4) and then minimizes the generalized entropy. Before the Page time, the quantum extremal surface that minimizes the generalized entropy shrinks to zero.

where $S_E[g]$ is the gravitational Euclidean action given by

$$S_E[g] = -\frac{1}{16\pi G}\int dx^d \sqrt{g}(R + \dots) + (\text{boundary terms}). \tag{10}$$

For simplicity, we will ignore higher curvature and boundary terms. Gravity is considered only near the classical saddle, and the saddle-point approximation gives $Z_{\text{GR}} \propto e^{-S_{\text{GR}}}$. To compute the entropy of fields on a gravitational background, we can instead consider its complementary region where no gravity effect is involved. To apply the replica trick, we require n copies of the original spacetime. In the gravitation regions, additional classical saddles may exist – most notably the "wormhole saddle" – on the n-cover manifold $\tilde{\mathcal{M}}_n$, as illustrated in **Figure 4**. While the manifold $\tilde{\mathcal{M}}_n$ is smooth everywhere, the orbifold defined by $\mathcal{M}_n = \tilde{\mathcal{M}}_n/\mathbb{Z}_n$ has conical singularities with an opening angle $2\pi/n$. The integration of the Ricci scalar in Eq. (10) can be done on the orbifold $\mathcal{M}_n$, which gives

$$\int dx^d \sqrt{g}R = 4\pi(1 - 1/n)\int_{\Sigma_{d-2}} \sqrt{g}, \tag{11}$$

where Σ_{d-2} is the codimension-2 slice. This integral can be interpreted as being performed on the cones near conical singularities [19]. Combining this with Eq. (6), we arrive at the generalized entropy formula:

$$S_{\text{gen}} = \frac{\text{Area}_{\text{cd=2}}}{4G_N} + S_{\text{matter}}, \tag{12}$$

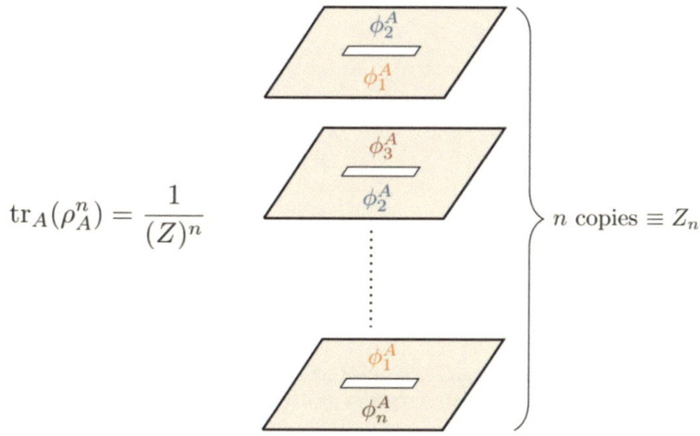

Figure 3.
Computing $\mathrm{tr}_A \rho_A^n$ *by gluing n copies of the original spacetime sheet with a cut along A. The gluing is represented by identifying the fields of the same colors. Figure from Ref. [44].*

where $\mathrm{Area}_{\mathrm{cd}=2}$ denotes the area of the codimension-2 surface, and S_{matter} represents the entropy of the CFT, which depends on the positions of the conical singularities. In fact, S_{matter} now includes contributions not only from the fields in the non-gravitational regions but also from inside the wormholes, i.e., between the conical singularities, due to the wormhole configurations. This result can be shown explicitly by inserting twister operators at conical singularities. As a consequence, this reproduces the semiclassical entropy $S_{\mathrm{semi-cl}(\Sigma_X)}$ as expressed in Eq. (4). Here, the cutoff surface corresponds to the boundaries of the non-gravitational regions. By extremizing Eq. (12), one can identify the quantum extremal surface.

This simple derivation already demonstrates how including contributions from Euclidean wormholes in the gravitational path integral can influence the calculation of entropy. The diagrammatic representation of replica wormholes and the orbifold $\mathcal{M}_n$ is illustrated in **Figure 4**. The configurations with wormholes are usually referred to as the "wormhole saddle," while those without wormholes are referred to as the "Hawking saddle." Generally, contributions from both saddles need to be evaluated to determine which dominates in the computation of entanglement entropy.

2.3 Entanglement entropy from the island formula

The gravitational entropy formula does not rely on the assumption of AdS/CFT correspondence and has been applied to the calculation of Page curves for black holes in spaces beyond asymptotically AdS, including both eternal and dynamic black holes. For more details, interested readers can refer to Refs. [25, 27, 45–55]. In this subsection, we will briefly discuss the results of these calculations and their implications.

The first step in obtaining the entropy of Hawking radiation is to compute the generalized entropy:

$$S_{\mathrm{gen}} = \frac{\mathrm{Area}(\partial I)}{4G_{\mathrm{N}}} + S_{\mathrm{matter}}(R \cup I), \tag{13}$$

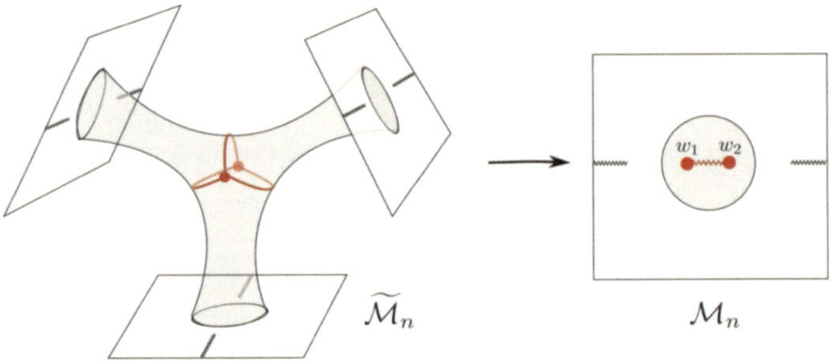

Figure 4.
An example of the covering manifold $\tilde{\mathcal{M}}_n$ and the orbifold $\mathcal{M}_n$ with $n = 3$. The non-gravitational regions where the entropy of quantum fields is calculated are represented by two black curvy lines. The conical singular points are at w_1 and w_2. Figure from Ref. [23].

where the second term corresponds to the entropy in the semi-classical approximation. After extremization, boundaries of the islands are found to either lie close to the event horizon or become trivial. For instance, the Penrose diagram of an island saddle for a static charged black hole is shown in **Figure 5**, where the island is the slice between the end points a_- and a_+ near the horizon.

For an island saddle, the first term in the generalized entropy resembles the Bekenstein-Hawking entropy of a black hole. Since the union $R \cup I$ occupies most of the spatial region on the Cauchy slice and the total entropy on the slice is zero, the generalized entropy can be well-approximated by the Bekenstein-Hawking entropy of the black hole. In the case of a trivial island, the generalized entropy reduces to the semi-classical entropy of the radiation. To obtain the von Neumann entropy, one has to select the minimal value between the Bekenstein-Hawking entropy and the semi-classical entropy of the radiation.

What Eq. (4) implies for an initially pure-state black hole is that the entropy of the emitted Hawking radiation is determined by the semi-classical entropy at the early stage of its evaporation. Prior to the Page time, the quantum extremal surface is trivial and Hawking's thermal radiation formula accurately computes the von Neumann entropy. During this phase, the entropy of Hawking radiation grows monotonically with time. However, after the Page time, the quantum extremal surface shifts to around the event horizon of the black hole. This region, isolated from the exterior radiation, is the entanglement island. Then, the dominant contribution to the entropy arises from the area of this extremal surface, which is approximately the Bekenstein-Hawking entropy. For an eternal two-sided black hole, the von Neumann entropy of Hawking radiation saturates at a constant value $2S_{BH}$ after the Page time, as illustrated in **Figure 5**. In contrast, for an evaporating black hole, the entropy decreases monotonically until the black hole has completely evaporated.

Interestingly, the Page curve derived from the island formula agrees with the result originally obtained by Page through pure quantum mechanical arguments [16, 17]. This agreement, combined with insights from the AdS/CFT correspondence, suggests that black holes can be treated as ordinary quantum systems with unitary dynamics. This perspective has spurred interest in integrating ideas and techniques from quantum information into the study of black holes.

3. Black holes as information mirrors

Black holes scramble quantum information rapidly [29, 30]. This feature motivates the representation of quantum dynamics inside a black hole by random unitary operations. Before the discovery of the quantum extremal surface, Hayden and Preskill used this approach to investigate when information escapes a black hole, assuming unitary dynamics inside a black hole [28]. They found that black holes essentially function as information mirrors, with infalling information being bounced back almost immediately upon reaching the event horizon. The discovery opens a new avenue for studying black holes from the quantum information perspective.

3.1 Hayden-Preskill thought experiment

In 2007, Hayden and Preskill demonstrated that information dropped into a black hole after the Page time can escape at an optimal rate, while information dropped in before the Page time can still escape shortly after the Page time. This finding was surprising, as the rapid recoverability of information was unexpected. Despite the astounding nature of this result, the argument and techniques behind it are remarkably straightforward, as we will explain in the following discussion.

One can imagine that Alice plans to destroy certain quantum information she holds by tossing her qubits M into a black hole. We assume that Alice's qubits M are maximally entangled with Charlie's qubits N, with $|M| = |N|$. The initial state of the quantum memory is the pure quantum state

$$|\Phi\rangle^{MN} = \frac{1}{\sqrt{N}} \sum_{a=1}^{|N|} |a\rangle^{M} \otimes |a\rangle^{N}. \tag{14}$$

Bob, who intends to decipher Alice's information from the Hawking radiation, has been collecting all the radiation emitted from the black hole since its formation. For an old black hole—one that has radiated away most of its initial mass—the system is nearly maximally entangled with its early radiation, denoted by E. Assuming Bob has full knowledge of the black hole's internal dynamics, which are modeled as a thorough scrambling of the infalling information via a known unitary transformation, an essential question arises: When will Bob be able to recover Alice's information from the Hawking radiation?

It is clear that when the whole black hole has evaporated unitarily, Bob will eventually retrieve Alice's information. The key question is about the earliest time the recovery becomes possible. The diagrammatic illustration of the setup is shown in **Figure 6**, where E is the early radiation that Bob has collected since the beginning of the black hole's formation, B is the black hole state after absorbing Alice's subsystem M, R is the new radiation emitted after the absorption of M, and V^B is the scrambling unitary chosen randomly with respect to the Haar measure on $U(|B|)$.

To determine whether Bob has successfully recovered the information that Alice attempted to destroy, we examine the entanglement between Charlie's system N and the remnant black hole B' as more and more radiation is emitted. If Bob can extract a subsystem from the Hawking radiation that is maximally entangled with N, we can conclude that he has recovered Alice's quantum memory. To begin, we calculate the distance between the system $B'N$ and the unentangled state $\sigma^N(V^B) \otimes \mathbb{I}^{B'}/\sqrt{|B'|}$,

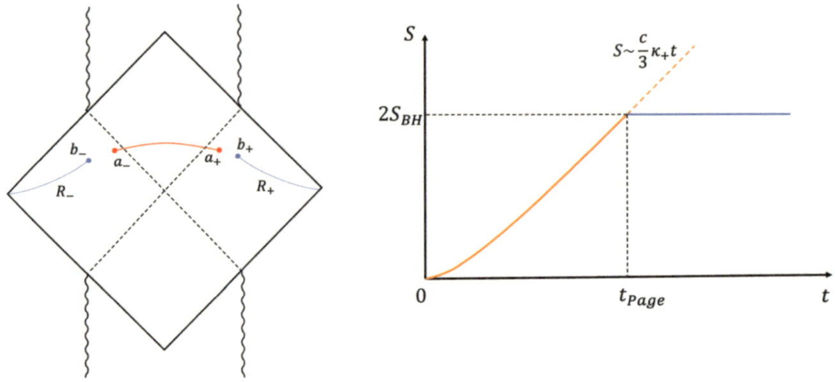

Figure 5.
An example of the island configuration (left) and the Page curve (right) for an eternal Reissner–Nordström black hole. The left figure shows the Penrose diagram of a charged black hole and its island after the Page time. The goal is to compute the entropy for the Hawking radiation on the two slices R_- and R_+. The island is represented by the red line between the ends a_- and a_+ near the horizon. On the right is the Page curve for the entropy. The dashed orange line shows the linear growth of entropy past the maximal entropy $2S_{BH}$ after the Page time t_{Page} according to the effective field calculation. The Island formula yields the blue solid line due to the dominance of the island saddle after t_{Page}. Figure from Ref. [54].

where $\sigma^N\left(V^B\right)$ is the reduced density matrix of system N after the scrambling operation V^B, defined as

$$\sigma(N) = \mathrm{Tr}_{RB'}\left[\left(\mathbb{I}^N \otimes V^B\right)\rho^{BN}\left(\mathbb{I}^N \otimes V^{B\dagger}\right)\right]. \tag{15}$$

Here, the system B refers to the states just before entering V^B, to which the scrambling is applied. The distance, averaged over the scrambling unitary V^B, can then be estimated as

$$d(NB') = \int dV^B \left\| \sigma^{NB'}\left(V^B\right) - \sigma^N\left(V^B\right) \otimes \mathbb{I}^{B'}/\sqrt{|B'|} \right\|_1^2 \leq \frac{|NB|}{|R|^2}\,\mathrm{Tr}\left[\left(\rho^{NB}\right)^2\right], \tag{16}$$

where the L^1 norm is defined by $|A|_1 = \mathrm{Tr}\sqrt{A^\dagger A}$. Since the system B is maximally entangled with NE, ρ_{NB} is maximally mixed in a system of dimension $|B|/|N|$, which gives

$$\mathrm{Tr}\left(\rho^{NB}\right)^2 = \frac{|N|}{|B|}. \tag{17}$$

Assuming that Alice has tossed k qubits into the black hole and r qubits have been radiated in the system R, the inequality Eq. (16) becomes

$$\int dV^B \left\| \sigma^{NB'}\left(V^B\right) - \sigma^N\left(V^B\right) \otimes \frac{\mathbb{I}^{B'}}{\sqrt{|B'|}} \right\|_1^2 \leq \frac{|N|^2}{|R|^2} = \frac{2^{2k}}{2^{2r}} = 2^{2(k-r)}. \tag{18}$$

This inequality shows that Charlie's qubits N will become nearly completely disentangled from the black hole B' as soon as only slightly more qubits in R are emitted from the black hole. When N becomes disentangled from the black hole, it will be in maximal entanglement with ER. Then, decoding maps can be designed with fidelity satisfying

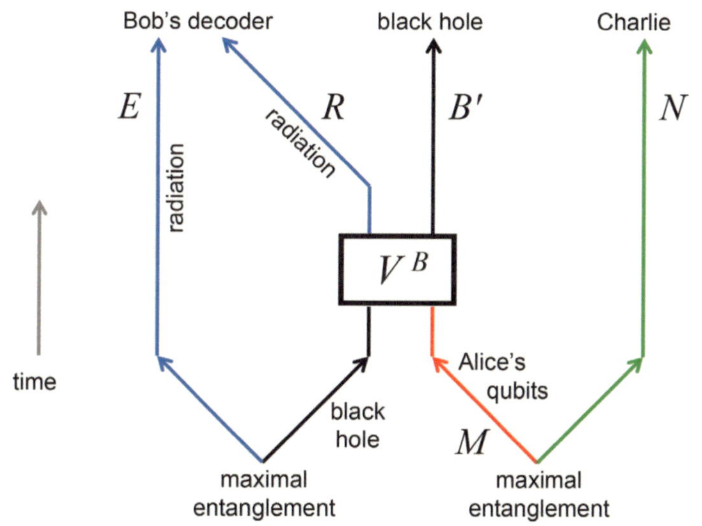

Figure 6.
Hayden-Preskill thought experiment on information retrieval from an evaporating black hole. Initially, the black hole is maximally entangled with its earlier radiation E, and Alice's qubits M are maximally entangled with Charlie's qubits N. After M is absorbed into the black hole, the total system B undergoes a random unitary scrambling V^B, after which new radiation R is emitted and the remnant black hole is represented by B'. Figure from Ref. [28].

$$F\left(V^B\right) \geq 1 - \left\| \sigma^{NB'}\left(V^B\right) - \sigma^N\left(V^B\right) \otimes \frac{\mathbb{I}^{B'}}{\sqrt{|B'|}} \right\|_1. \tag{19}$$

Averaging over the unitary V^B, the fidelity satisfies $F \geq 1 - 2^{-(k-r)}$. Therefore, Bob will be very confident that he can successfully recover the information Alice tried to destroy shortly after he has collected a few more qubits than Alice tossed.

From the calculations above, we can readily conclude that if Alice dropped k qubits into the black hole after the Page time, Bob only needs to wait until the black hole has emitted approximately $\mathcal{O}(k)$ qubits to recover the information with appreciable confidence. Apparently, the time it takes is negligible compared to the Page time or the lifetime of the black hole; thus, a black hole essentially functions as an information mirror, reflecting back infalling information almost as soon as it enters the black hole.

Moreover, the above result has a correspondence in the quantum extremal surface formulation. From the calculation of entanglement entropy, it is known that the boundary of an entanglement island after Page time lies approximately on the event horizon. Therefore, if we toss information into the black hole, it will not significantly alter the island structure, assuming its energy is much less than the black hole's mass. As information inside quantum extremal surfaces is interpreted as effectively encoded in the radiation via entanglement wedge reconstruction [26, 56, 57], analogous to the holographic screen of the RT surface in **Figure 1**, the information can be retrieved almost immediately after reaching the horizon. Remarkably, this reasoning from the gravity side aligns with the quantum information argument presented above, where black hole dynamics are modeled as a random unitary.

3.2 Information decoding from black holes

In the previous subsection, we have seen that information dropped into a black hole after the Page time can, in principle, be quickly decoded from the Hawking radiation. In this subsection, we will show explicitly how to decode the information starting with the decoding protocol proposed by Yoshida and Kitaev [31]. With an explicit decoding protocol, one may simulate the information recovery from black hole radiation by running the algorithm on quantum computers and testing it in practice!

3.2.1 Yoshida-Kitaev protocol

The Yoshida-Kitaev decoding protocol is based on the Hayden-Preskill thought experiment. The idea is to prepare a similar auxiliary system and perform a corresponding scrambling unitary, such that the information initially stored in the entanglement between R and A in **Figure** 7 is shifted to the entanglement between R and a corresponding output state of the same dimension, R'.

On the left panel of **Figure** 7, the Hayden-Preskill thought experiment is illustrated. Alice's qubits, which are maximally entangled with R, are isometrically embedded into the quantum state A. Here, $|\xi\rangle = (I_R \otimes \Xi_\alpha)|\mathrm{EPR}\rangle_{R\alpha}$, where $\Xi : \alpha \to A$ is an isometric embedding of Alice's qubits α. The outputs of the random unitary U are the new radiation D, represented by the red arrow, and the remnant black hole C. The goal for Bob is to perform an operation V on subsystems D and B' to recover the EPR state $|\mathrm{EPR}\rangle_{RR'}$. The explicit decoding algorithm is shown in the right panel of **Figure** 7. The operations in the circled shaded region correspond to the operation V shown in the left panel.

In this decoding protocol, Bob has to measure the output state on the joint subsystem $|DD'\rangle$. Assuming U is a Haar-random unitary gate, Bob can only successfully decode the information if the projection onto $|\mathrm{EPR}\rangle_{DD'}$ is realized. However, the probability of a successful EPR projection is

$$P_{\mathrm{EPR}} \approx \frac{1}{d_A d_R} + \frac{1}{d_D^2} \leq \frac{1}{d_R^2}, \tag{20}$$

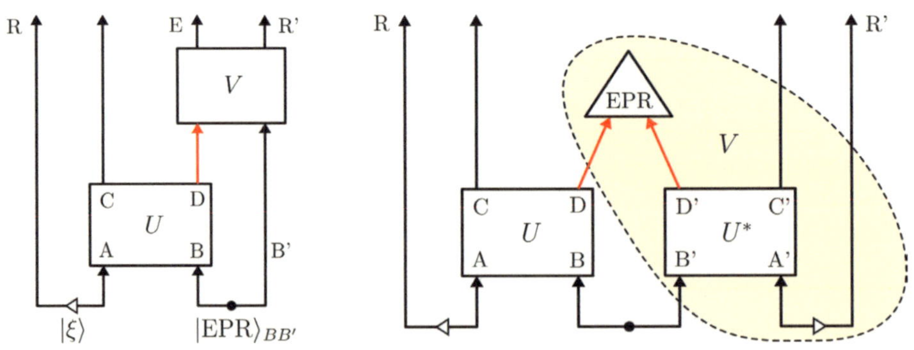

Figure 7.
The Yoshida-Kitaev decoding protocol for executing the information retrieval procedure from an evaporating black hole. The Hayden-Preskill thought experiment is shown in the left panel. Initially, the black hole B and its Hawking radiation B′ are maximally entangled. Alice's qubits, which are maximally entangled with R, are embedded isometrically into the quantum state A. After the random unitary U, the new radiation D is emitted, represented by the red arrow, and the remnant black hole C remains. Bob will perform an operation V on the systems D and B′ to recover the information. In the right panel, an explicit algorithm is presented. Bob will prepare primed systems and project D and D′ onto the EPR state. If the projection is successful, Bob will successfully decode the information, and the subsystems R′ and R will form an EPR state. Figure from Ref. [31].

which means that even if enough radiation is collected by Bob, the chance of successful decoding will not be much better than pure guessing, except that he will know when he gets the information correctly. Fortunately, this problem of selecting a particular state is a well-known issue in quantum information theory and can be solved by methods such as Grover's search algorithm [58]. By implementing such search algorithms, the EPR projections can be replaced by the search protocols, and the problem of a low success probability can be avoided. Therefore, one could decode the information with significantly higher confidence.

3.2.2 Horowitz-Maldacena model and decoding in effective descriptions

The decoding protocol discussed in the previous subsection is implemented on the modes in the boundary theory, where effective modes inside the black hole are excluded. In this description, no information paradox is present, and the entangled Hawking pairs generated near the horizon are not considered. However, in the effective field theory, black holes emit mixed-state radiation entangled with partner modes inside the horizon. This effective description, originally thought to be valid in the weak gravity regime, contrasts with the unitary description from boundary theories. The Horowitz-Maldacena (HM) proposal [59] attempts reconciliation by imposing a final-state boundary condition at the singularity, enabling information escape via quantum teleportation. However, it faces challenges, including causality violations and inconsistency with the Page curve [60, 61]. Recent refinements reinterpret information transfer as non-isometric encoding. By including a scrambling unitary followed by a nonunitary projection onto a fixed state, one can show that this simple picture can reproduce the Island formula initially obtained through the gravitational path integral. The reason for its success may be related to the fact that the Haar average in a finite-dimensional Hilbert space can be seen as a projection into a subspace P of the replica Hilbert space used for computing Renyi entropies [62, 63]. Moreover, decoding protocols similar to the Yoshida-Kitaev protocol have been applied to the final-state projection model to test information recoverability in the model [36]. In contrast to the original Hayden-Preskill thought experiment, where only unitary dynamics are present, whether information is still conserved when a large portion of modes are reduced through the non-unitary projection process is no longer obvious. The information conservation in the modified final-state projection model has been explicitly shown, and the decoding protocols have been tested on quantum computers.

The diagrammatic illustration of the decoding strategy is shown in **Figure 8**. The black hole is formed from a pure state $|\psi_0\rangle_f$. A and A' represent the subsystem Alice tossed into the black hole and its entangled counterpart, respectively. r denotes the infalling modes inside the black hole, while R represents the Hawking radiation maximally entangled with r. After the application of a random unitary U—which can also be interpreted as a mapping to the fundamental picture—redundant modes must be projected onto a fixed state using operators such as $\langle 0|_P$ to conserve entropy. Finally, B and R' represent the remnant black hole and the new radiation, respectively.

Compared to the original Hayden-Preskill protocol, the ingoing modes r and the projection $\langle 0|_P$ are explicitly included in this framework. Interestingly, for black holes that have existed longer than their Page time, the EPR projections as required in the Yoshida-Kitaev protocol are no longer necessary in this model

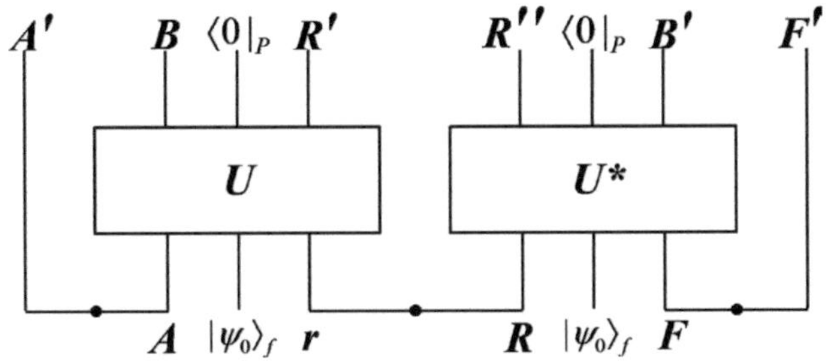

Figure 8.
The Hayden-Preskill thought experiment for the modified final-state projection model. The model is presented in the left half of the diagram, and the right half shows the decoder. The black hole is formed by the pure state $|\psi_0\rangle_f$. A and A′ represent the subsystem Alice tossed into the black hole and its entangled subsystem, respectively. r represents the infalling modes inside the black hole, while R represents the Hawking radiation that is maximally entangled with r. After the random unitary U, which is also viewed as the mapping to the fundamental picture, redundant modes must be projected onto a fixed state by operators such as $\langle 0|_P$ to conserve entropy. B and R′ represent the remnant black hole and the new radiation, respectively. Figure from Ref. [38].

due to the annihilation of the majority of modes inside the black hole by the local projections.

4. Conclusions

The progress in understanding black hole information is a monumental achievement in theoretical physics, combining insights from quantum field theories, gravity, and quantum computing. The discovery of the Page curve—and its agreement with the principles of unitary quantum evolution—supports that information is conserved during black hole evaporation. The Island formula not only provides a computational tool of entropy but also reveals a deeper entanglement structure underlying the late-time behavior of black holes. Its results align with results derived from quantum information arguments: information is not lost inside black holes but delicately encoded in the Hawking radiation and released shortly after the Page time. This synthesis of gravitational and quantum information perspectives not only helps resolve the long-standing paradox but also opens new avenues for understanding the fundamental nature of spacetime and quantum gravity.

In this chapter, we explored the connections between gravity, quantum mechanics, and quantum information theory in addressing black hole information problems. Beginning with the Ryu-Takayanagi formula, we derived the Island formula from gravitational path integrals, offering a generalized framework for computing the entanglement entropy of black holes and their radiation. We demonstrated its effectiveness in calculating the entropy of Hawking radiation and explored its implications for both static and evaporating black holes. Moving beyond gravitational formulation, we examined how concepts and tools from quantum information theory, exemplified by the Hayden-Preskill thought experiment and Yoshida-Kitaev algorithm, shed light on how information can be extracted from Hawking radiation in simple quantum qubit models. We also applied the decoding protocol to the final-state projection models, showing that information is conserved even when states falling into the black hole are projected nonunitarily onto local states, thus extending the scope of effectively unitary descriptions. These studies also provide a quantum algorithm approach for simulating the information recovery from black holes on quantum computers.

In addition to the gravitational path integral and quantum information formalisms discussed in the previous subsections, many other qubit models have been proposed as alternative and effective descriptions of black hole entanglement dynamics. For a non-exhaustive list, see [37, 59, 64–69]. For example, Wang et al. [37] explores potential connections between Page transitions in the entanglement entropy and quantum-to-classical transitions in quantum measurements, where the state purification in the late stage of Hawking radiation is interpreted as an effective measurement. In [64], non-local gravitational degrees of freedom are proposed as a mechanism for allowing information to escape from the black hole.

Despite these advances, several questions remain. Though the generalized entropy formula is consistent with unitarity, the microscopic mechanism behind these results is not entirely transparent [70–72]. Questions on the physical meaning of the Euclidean wormholes, the smoothness of horizons, and the exact quantum state of Hawking radiation are examples where the Island formula has not provided fully satisfying explanations. Furthermore, since the Island formula corroborates Hawking's effective field calculations before the Page time, it is unclear why Hawking's radiation formula fails afterward. The avoidance of firewalls within this framework is also not entirely transparent.

The collaboration between black hole physics and quantum information science has yielded fruitful results for both fields. As technology advances and theoretical understanding continues to evolve, future investigations may bring more surprising insights into the quantum nature of black holes.

Acknowledgement

The author acknowledges the use of ChatGPT-4o mini for language polishing of the manuscript.

Author details

Xuanhua Wang
Wenzhou Institute, University of Chinese Academy of Sciences, Wenzhou, Zhejiang, China

*Address all correspondence to: wang.nihao.phys@gmail.com

References

[1] Giddings SB. Quantum Mechanics of Black Holes [arXiv:hep-th/9412138 [hep-th]]

[2] Aharony O, Gubser SS, Maldacena JM, Ooguri H, Oz Y. Large N field theories, string theory and gravity. Physics Reports. 2000;**323**:183-386. DOI: 10.1016/S0370-1573(99)00083-6 [arXiv: hep-th/9905111 [hep-th]]

[3] Harlow D. Jerusalem lectures on black holes and quantum information. Reviews of Modern Physics. 2016;**88**: 015002 [arXiv:1409.1231 [hep-th]]

[4] Peres A, Terno DR. Quantum information and relativity theory. Reviews of Modern Physics. 2004;**76**:93-123. DOI: 10.1103/RevModPhys.76.93 [arXiv:quant-ph/0212023 [quant-ph]]

[5] Hawking SW. Particle creation by black holes. Communications in Mathematical Physics. 1975;**43**:199 Erratum: [Commun. Math. Phys. 46, 206 (1976)]

[6] Wald RM. On particle creation by black holes. Communications in Mathematical Physics. 1975;**45**:9-34

[7] Parker L. Probability distribution of particles created by a black hole. Physical Review D. 1975;**12**:1519-1525

[8] Almheiri A, Marolf D, Polchinski J, Sully J. Black holes: Complementarity or firewalls? Journal of High Energy Physics. 2013;**2**:062 [arXiv:1207.3123 [hep-th]]

[9] Almheiri A, Marolf D, Polchinski J, Stanford D, Sully J. An apologia for firewalls. Journal of High Energy Physics. 2013;**09**:018. DOI: 10.1007/JHEP09(2013)018 [arXiv:1304.6483 [hep-th]]

[10] Hawking SW. Breakdown of predictability in gravitational collapse. Physical Review D. 1976;**14**:2460-2473. DOI: 10.1103/PhysRevD.14.2460

[11] Mathur SD. The information paradox: A pedagogical introduction. Classical and Quantum Gravity. 2009; **26**:224001. DOI: 10.1088/0264-9381/26/22/224001 [arXiv:0909.1038 [hep-th]]

[12] Giddings SB. Black hole information, unitarity, and nonlocality. Physical Review D. 2006;**74**:106005. DOI: 10.1103/PhysRevD.74.106005 [arXiv: hep-th/0605196 [hep-th]]

[13] Bekenstein JD. Black holes and entropy. Physical Review D. 1973;**7**:2333-2346. DOI: 10.1103/PhysRevD.7.2333

[14] Bekenstein JD. A universal upper bound on the entropy to energy ratio for bounded systems. Physical Review D. 1981;**23**:287

[15] Bousso R. The holographic principle. Reviews of Modern Physics. 2002;**74**: 825-874. DOI: 10.1103/RevModPhys.74.825 [arXiv:hep-th/0203101 [hep-th]]

[16] Page DN. Information in black hole radiation. Physical Review Letters. 1993; **71**:3743-3746 [arXiv:hep-th/9306083 [hep-th]]

[17] Page DN. Average entropy of a subsystem. Physical Review Letters. 1993;**71**:1291-1294 [arXiv:gr-qc/9305007 [gr-qc]]

[18] Gibbons GW, Shellard EPS. The future of theoretical physics and cosmology: Celebrating Stephen Hawking's 60th birthday. In: Proceedings, Workshop and

Symposium; 7-10 January 2002, Cambridge, UK; Cambridge, UK: Cambridge University Press; 2003

[19] Lewkowycz A, Maldacena J. Generalized gravitational entropy. Journal of High Energy Physics. 2013;**08**: 090. DOI: 10.1007/JHEP08(2013)090 [arXiv:1304.4926 [hep-th]]

[20] Engelhardt N, Wall AC. Extremal surface barriers. Journal of High Energy Physics. 2014;**03**:068. DOI: 10.1007/ JHEP03(2014)068 [arXiv:1312.3699 [hep-th]]

[21] Engelhardt N, Wall AC. Quantum extremal surfaces: Holographic entanglement entropy beyond the classical regime. Journal of High Energy Physics. 2015;**01**:073. DOI: 10.1007/ JHEP01(2015)073 [arXiv:1408.3203 [hep-th]]

[22] Almheiri A, Mahajan R, Maldacena J, Zhao Y. The page curve of Hawking radiation from semiclassical geometry. Journal of High Energy Physics. 2020;**03**: 149. DOI: 10.1007/JHEP03(2020)149 [arXiv:1908.10996 [hep-th]]

[23] Almheiri A, Hartman T, Maldacena J, Shaghoulian E, Tajdini A. Replica wormholes and the entropy of Hawking radiation. Journal of High Energy Physics. 2020;**05**:013. DOI: 10.1007/ JHEP05(2020)013 [arXiv:1911.12333 [hep-th]]

[24] Penington G, Shenker SH, Stanford D, Yang Z. Replica wormholes and the black hole interior. Journal of High Energy Physics. 2022;**03**:205. DOI: 10.1007/JHEP03(2022)205 [arXiv: 1911.11977 [hep-th]]

[25] Almheiri A, Engelhardt N, Marolf D, Maxfield H. The entropy of bulk quantum fields and the entanglement wedge of an evaporating black hole. Journal of High Energy Physics. 2019;**12**: 063. DOI: 10.1007/JHEP12(2019)063 [arXiv:1905.08762 [hep-th]]

[26] Penington G. Entanglement wedge reconstruction and the information paradox. Journal of High Energy Physics. 2020;**09**:002. DOI: 10.1007/JHEP09 (2020)002 [arXiv:1905.08255 [hep-th]]

[27] Almheiri A, Hartman T, Maldacena J, Shaghoulian E, Tajdini A. The entropy of Hawking radiation. Reviews of Modern Physics. 2021;**93**(3):035002. DOI: 10.1103/RevModPhys.93.035002 [arXiv:2006.06872 [hep-th]]

[28] Hayden P, Preskill J. Black holes as mirrors: Quantum information in random subsystems. Journal of High Energy Physics. 2007;**09**:120. DOI: 10.1088/1126-6708/2007/09/120 [arXiv: 0708.4025 [hep-th]]

[29] Sekino Y, Susskind L. Fast scramblers. Journal of High Energy Physics. 2008;**10**:065. DOI: 10.1088/ 1126-6708/2008/10/065 [arXiv: 0808.2096 [hep-th]]

[30] Maldacena J, Shenker SH, Stanford D. A bound on chaos. Journal of High Energy Physics. 2016;**08**:106. DOI: 10.1007/JHEP08(2016)106 [arXiv: 1503.01409 [hep-th]]

[31] Yoshida B, Kitaev A. Efficient decoding for the Hayden-Preskill protocol [arXiv:1710.03363 [hep-th]]

[32] Yoshida B, Yao NY. Disentangling scrambling and Decoherence via quantum teleportation. Physical Review X. 2019;**9**(1):011006. DOI: 10.1103/ PhysRevX.9.011006 [arXiv:1803.10772 [quant-ph]]

[33] Landsman KA, Figgatt C, Schuster T, Linke NM, Yoshida B, Yao NY, et al. Verified quantum information

scrambling. Nature. 2019;**567**(7746):61-65. DOI: 10.1038/s41586-019-0952-6 [arXiv:1806.02807 [quant-ph]]

[34] Brown AR, Gharibyan H, Penington G, Susskind L. The Python's lunch: Geometric obstructions to decoding Hawking radiation. Journal of High Energy Physics. 2020;**08**:121. DOI: 10.1007/JHEP08(2020)121 [arXiv: 1912.00228 [hep-th]]

[35] Bao N, Kikuchi Y. Hayden-Preskill decoding from noisy Hawking radiation. Journal of High Energy Physics. 2021;**02**:017. DOI: 10.1007/JHEP02(2021)017 [arXiv:2009.13493 [quant-ph]]

[36] Li R, Wang J. Hayden-Preskill protocol and decoding Hawking radiation at finite temperature. Physical Review D. 2022;**106**(4):046011. DOI: 10.1103/PhysRevD.106.046011 [arXiv: 2108.09144 [hep-th]]

[37] Wang X, Zhang K, Wang J. Entanglement islands, fire walls and state paradox from quantum teleportation and entanglement swapping. Classical and Quantum Gravity. 2023;**40**(9):095012. DOI: 10.1088/1361-6382/acc829 [arXiv: 2107.09228 [hep-th]]

[38] Li R, Wang X, Zhang K, Wang J. Information retrieval from Hawking radiation in the non-isometric model of black hole interior: Theory and quantum simulation. Physical Review D. 2024;**109**(4):044005. DOI: 10.1103/PhysRevD.109.044005 [arXiv: 2307.01454 [hep-th]]

[39] Nielsen MA, Chuang IL. Quantum Computation and Quantum Information: 10th Anniversary Edition. Cambridge, UK: Cambridge University Press; 2010. ISBN 978-0-521-63503-5. DOI: 10.1017/cbo9780511976667

[40] Hubeny VE, Rangamani M, Takayanagi T. A covariant holographic entanglement entropy proposal. Journal of High Energy Physics. 2007;**07**:062. DOI: 10.1088/1126-6708/2007/07/062 [arXiv:0705.0016 [hep-th]]

[41] Ryu S, Takayanagi T. Aspects of holographic entanglement entropy. Journal of High Energy Physics. 2006;**08**:045. DOI: 10.1088/1126-6708/2006/08/045 [arXiv:hep-th/0605073 [hep-th]]

[42] Faulkner T, Lewkowycz A, Maldacena J. Quantum corrections to holographic entanglement entropy. Journal of High Energy Physics. 2013;**11**:074. DOI: 10.1007/JHEP11(2013)074 [arXiv:1307.2892 [hep-th]]

[43] Ryu S, Takayanagi T. Holographic derivation of entanglement entropy from AdS/CFT. Physical Review Letters. 2006;**96**:181602. DOI: 10.1103/PhysRevLett.96.181602 [arXiv:hep-th/0603001 [hep-th]]

[44] Nishioka T. Entanglement entropy: Holography and renormalization group. Reviews of Modern Physics. 2018;**90**(3):035007. DOI: 10.1103/RevModPhys.90.035007 [arXiv: 1801.10352 [hep-th]]

[45] Hollowood TJ, Kumar SP. Islands and Page curves for evaporating black holes in JT gravity. Journal of High Energy Physics. 2020;**08**:094. DOI: 10.1007/JHEP08(2020)094 [arXiv: 2004.14944 [hep-th]]

[46] Balasubramanian V, Kar A, Ugajin T. Islands in de Sitter space. Journal of High Energy Physics. 2021;**02**:072. DOI: 10.1007/JHEP02(2021)072 [arXiv: 2008.05275 [hep-th]]

[47] Gautason FF, Schneiderbauer L, Sybesma W, Thorlacius L. Page curve for an evaporating black hole. Journal of

High Energy Physics. 2020;**05**:091. DOI: 10.1007/JHEP05(2020)091 [arXiv: 2004.00598 [hep-th]]

[48] Hartman T, Shaghoulian E, Strominger A. Islands in asymptotically flat 2D gravity. Journal of High Energy Physics. 2020;**07**:022. DOI: 10.1007/JHEP07(2020)022 [arXiv:2004.13857 [hep-th]]

[49] Miyata A, Ugajin T. Evaporation of black holes in flat space entangled with an auxiliary universe. Progress of Theoretical and Experimental Physics. 2022;**2022**(1):013B13. DOI: 10.1093/ptep/ptab163 [arXiv:2104.00183 [hep-th]]

[50] Balasubramanian V, Kar A, Ugajin T. Entanglement between two disjoint universes. Journal of High Energy Physics. 2021;**02**:136. DOI: 10.1007/JHEP02(2021)136 [arXiv:2008.05274 [hep-th]]

[51] Hartman T, Jiang Y, Shaghoulian E. Islands in cosmology. Journal of High Energy Physics. 2020;**11**:111. DOI: 10.1007/JHEP11(2020)111 [arXiv: 2008.01022 [hep-th]]

[52] Geng H, Nomura Y, Sun HY. Information paradox and its resolution in de sitter holography. Physical Review D. 2021;**103**(12):126004. DOI: 10.1103/PhysRevD.103.126004 [arXiv: 2103.07477 [hep-th]]

[53] Marolf D, Maxfield H. Observations of Hawking radiation: The Page curve and baby universes. Journal of High Energy Physics. 2021;**04**:272. DOI: 10.1007/JHEP04(2021)272 [arXiv: 2010.06602 [hep-th]]

[54] Wang X, Li R, Wang J. Islands and page curves of Reissner-Nordström black holes. Journal of High Energy Physics. 2021;**04**:103. DOI: 10.1007/JHEP04(2021)103 [arXiv:2101.06867 [hep-th]]

[55] Wang X, Li R, Wang J. Page curves for a family of exactly solvable evaporating black holes. Physical Review D. 2021;**103**(12):126026. DOI: 10.1103/PhysRevD.103.126026 [arXiv: 2104.00224 [hep-th]]

[56] Pastawski F, Yoshida B, Harlow D, Preskill J. Holographic quantum error-correcting codes: Toy models for the bulk/boundary correspondence. Journal of High Energy Physics. 2015;**06**:149. DOI: 10.1007/JHEP06(2015)149 [arXiv: 1503.06237 [hep-th]]

[57] Cotler J, Hayden P, Penington G, Salton G, Swingle B, Walter M. Entanglement wedge reconstruction via universal recovery channels. Physical Review X. 2019;**9**(3):031011. DOI: 10.1103/PhysRevX.9.031011 [arXiv: 1704.05839 [hep-th]]

[58] Grover LK. A fast quantum mechanical algorithm for database search. In: Proceedings of the Twenty-Eighth Annual ACM Symposium on Theory of Computing, STOC'96. New York, NY, USA: ACM; 1996. pp. 212-219. DOI: 10.1145/237814.237866

[59] Horowitz GT, Maldacena JM. The black hole final state. Journal of High Energy Physics. 2004;**02**:008. DOI: 10.1088/1126-6708/2004/02/008 [arXiv: hep-th/0310281 [hep-th]]

[60] Lloyd S, Preskill J. Unitarity of black hole evaporation in final-state projection models. Journal of High Energy Physics. 2014;**08**:126. DOI: 10.1007/JHEP08(2014)126 [arXiv:1308.4209 [hep-th]]

[61] Gottesman D, Preskill J. Comment on 'the black hole final state'. Journal of High Energy Physics. 2004;**03**:026. DOI:

10.1088/1126-6708/2004/03/026 [arXiv: hep-th/0311269 [hep-th]]

[62] Zhou T, Nahum A. Entanglement membrane in chaotic many-body systems. Physics Review X. 2020;**10**(3): 031066. DOI: 10.1103/PhysRevX.10.031066 [arXiv:1912.12311 [cond-mat.str-el]]

[63] Liu H, Vardhan S. Entanglement entropies of equilibrated pure states in quantum many-body systems and gravity. PRX Quantum. 2021;**2**(1): 010344. DOI: 10.1103/PRXQuantum.2.010344 [arXiv: 2008.01089 [hep-th]]

[64] Osuga K, Page DN. Qubit transport model for unitary black hole evaporation without firewalls. Physical Review D. 2018;**97**(6):066023. DOI: 10.1103/PhysRevD.97.066023 [arXiv:1607.04642 [hep-th]]

[65] Chatwin-Davies A, Jermyn AS, Carroll SM. How to recover a qubit that has fallen into a black hole. Physical Review Letters. 2015;**115**(26):261302. DOI: 10.1103/PhysRevLett.115.261302 [arXiv:1507.03592 [hep-th]]

[66] Borsten L, Duff MJ, Levay P. The black-hole/qubit correspondence: An up-to-date review. Classical and Quantum Gravity. 2012;**29**:224008. DOI: 10.1088/0264-9381/29/22/224008 [arXiv: 1206.3166 [hep-th]]

[67] Bousso R. Violations of the equivalence principle by a nonlocally reconstructed vacuum at the black hole horizon. Physical Review Letters. 2014; **112**(4):041102. DOI: 10.1103/PhysRevLett.112.041102 [arXiv: 1308.3697 [hep-th]]

[68] Papadodimas K, Raju S. An Infalling Observer in AdS/CFT. Journal of High Energy Physics. 2013;**10**:212. DOI: 10.1007/JHEP10(2013)212 [arXiv: 1211.6767 [hep-th]]

[69] Harlow D. Aspects of the Papadodimas-Raju proposal for the black hole interior. Journal of High Energy Physics. 2014;**11**:055. DOI: 10.1007/JHEP11(2014)055 [arXiv:1405.1995 [hep-th]]

[70] Bousso R, Tomašević M. Unitarity from a smooth horizon? Physical Review D. 2020;**102**(10):106019. DOI: 10.1103/PhysRevD.102.106019 [arXiv: 1911.06305 [hep-th]]

[71] Bousso R, Wildenhain E. Gravity/ensemble duality. Physical Review D. 2020;**102**(6):066005. DOI: 10.1103/PhysRevD.102.066005 [arXiv: 2006.16289 [hep-th]]

[72] Geng H, Karch A, Perez-Pardavila C, Raju S, Randall L, Riojas M, et al. Inconsistency of islands in theories with long-range gravity. Journal of High Energy Physics. 2022;**01**:182. DOI: 10.1007/JHEP01(2022)182 [arXiv: 2107.03390 [hep-th]]